Inverse Problems

SIAM
AMS
proceedings

volume 14

Inverse Problems

AMERICAN MATHEMATICAL SOCIETY
PROVIDENCE · RHODE ISLAND

PROCEEDINGS OF THE SYMPOSIUM IN APPLIED MATHEMATICS
OF THE AMERICAN MATHEMATICAL SOCIETY
AND THE SOCIETY FOR INDUSTRIAL AND APPLIED MATHEMATICS

HELD IN NEW YORK
APRIL 12–13, 1983

EDITED BY
D. W. McLAUGHLIN

Prepared by the American Mathematical Society with support from
NSF grant MCS 8219934

1980 *Mathematics Subject Classification.* Primary 34A55, 34B25, 34F05, 35P25.

Library of Congress Cataloging in Publication Data

Symposium in Applied Mathematics (1983: New York, N.Y.)
Inverse problems.

(SIAM-AMS proceedings, ISSN 0080-5084; v. 14)

"Proceedings of the Symposium in Applied Mathematics of the American Mathematical Society and the Society for Industrial and Applied Mathematics, held in New York City, April 12–13, 1983"–T.p. verso.

Bibliography: p.

1. Inverse problems (Differential equations)–Congresses. I. McLaughlin, D. W. (David W.), 1944– . II. American Mathematical Society. III. Society for Industrial and Applied Mathematics. IV. Title. V. Series.

QA370.S96 1983 515.3′5 84-392
ISBN 0-8218-1334-X (alk. paper)

TABLE OF CONTENTS

PREFACE

This volume contains the proceedings of a symposium on inverse methods which was held on April 12 and 13, 1983, in New York City as a part of the regional meeting of the American Mathematical Society. The organizing committee for the symposium consisted of Robert Burridge, New York University; Joseph B. Keller, Stanford University; R. B. Marr, Brookhaven National Laboratory; David W. McLaughlin (Chairman), University of Arizona; C. R. Smith, University of Wyoming. Our goal in organizing the conference was to illustrate the breadth of modern inverse problems, both with regard to the diversity of applications and the diversity of mathematical methods. Because the field is so broad, we had to be content with representative areas. The conference consisted in four half-day sessions on the following topics: (i) geophysical inverse problems, (ii) computer tomography and inverse problems in medicine, (iii) developments in mathematical inverse theory; (iv) methods of maximum information entropy. The ordering of papers in this volume is the same as the ordering of presentations at the meeting.

Finally, I certainly want to thank the staff of the American Mathematical Society, and in particular Ellen Heiser, Dorothy Smith, and Tony Palermino whose technical assistance made the organization of the symposium and the preparation of these proceedings extremely easy.

David W. McLaughlin
University of Arizona

Part I. Geophysical Inverse Problems

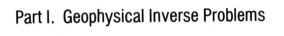

SIAM-AMS PROCEEDINGS
Volume 14
1984

An Inverse Problem of Electromagnetism Arising in Geophysics

ROBERT L. PARKER[1]

ABSTRACT. Magnetotelluric data routinely collected in geophysical surveys consist of measurements of the ratio of electric and perpendicular magnetic fields at surface of the Earth. The ratio is effectively an impedance and is obtained as a function of frequency. Under the simplification that the underground structure is one-dimensional, the problem of finding the electrical conductivity from the data can be treated as an inverse scattering problem and solved by a number of long-established techniques (e.g. that of Gel'fand and Levitan). It is important, however, with practical data to avoid the idealizations of classical theories; specifically, one must acknowledge the finiteness of the number of observations and their imprecision. Much is gained by representing the observed data as linear functionals of the spectral function of the differential equation for the system. Then the question of existence can be solved in terms of a semi-infinite linear or quadratic program. If any solution exists, there are many of them. To construct a particular conductivity profile compatible with the data necessitates some sort of selection from the mainfold of admissible models. Three special classes are considered in which construction is efficient and relatively stable numerically. The matter of making useful inferences about the general properties of all admissible solutions is much more difficult. One negative result exists: there is a depth below which the conductivity function may be chosen arbitrarily without destroying the agreement between the observations and the admittance of the model.

1. Introduction. At any fixed point on the Earth's surface the electromagnetic field is observed to be continually fluctuating. On time scales longer than about a year, it has been established that the cause of the variations is inside the Earth and is probably due to redistribution of the electric currents in the core that give rise to the global magnetic field. The source of the variations with shorter time scales is outside the Earth, either in the upper atmosphere, in the particle belts of the magnetosphere or even further out in

1980 *Mathematics Subject Classification.* Primary 86-A25
Key words and phrases. Inverse problems, scattering theory.
[1]Supported by National Science Foundation Grant EAR-79-19467.

space. Geophysicists use the electric and magnetic fields of exterior origin in a kind of giant scattering experiment to learn about the electrical conductivity within the Earth. If a sinusoidal field impinges on a uniformly conducting half space, there is an exponential decay of the fields in the conducting medium with a characteristic length scale, called the skin depth, which is inversely proportional to the square root of the frequency of the incident energy. Thus by studying slowly varying fields with periods of days to months we can learn about the conductivities at great depths (down to 1000 km), while with more rapidly changing fields we study the regions much nearer the surface; in fact, there is considerable commercial interest in the magnetotelluric method as a means of discovering mineral and hydrocarbon deposits with depths ranging from hundreds of meters to several kilometers.

For simplicity let us suppose the conductivity varies only with depth (or radius in the very large scale problems) and the source fields have long wavelengths compared to typical depths of interest. Then the mathematical question of recoving the conductivity can be transformed precisely into an inverse scattering problem of a classical kind. It might be thought that, because such problems have been understood rather well for nearly thirty years, this geophysical variant would be without mathematical interest. However, the nature of the actual observations and the ill-posed character of the inverse problem require a careful study of some aspects which, as far as I know, have been largely neglected in the scattering theory literature.

Before I get into the mathematical questions let me describe briefly how actual field observations are collected and analyzed, because, as I mentioned, it has an important bearing on the correct approach to the theory. For a comprehensive treatment the text by Rokityansky [**1982**] is recommended. In a typical magnetotelluric study horizontal electric and magnetic fields are recorded simultaneously at an observing station; in a typical commercial survey, samples are recorded at a rate of several thousand times per second for periods of a day or so. As we shall see, the quantity that is diagnostic of the interior conductivity is the ratio of the horizontal electric field and the perpendicular horizontal magnetic field considered as a function of frequency. In reality both field quantities are contaminated by unwanted noise — man-made and natural electric and magnetic fields of local origin or instrumental noise in the electrodes or electronics may all contribute because the fields we seek are quite tiny (a few microvolts per meter in electric field and a nanotelsa or less in magnetic field; both constitute less than a part in ten thousand of typical ambient steady values). In order to obtain useful information the measured values are cross-correlated to extract the coherent signals; this signal processing technique results in estimates of the required ratio at a finite number of frequencies and each estimate is associated with a statistical assessment of accuracy. For surveys concerned with shallow regions thirty independent frequencies are typically estimated, covering a band of three or more decades; in long-period studies the number of frequencies at which estimates are made may be smaller (10 or 15) and usually the range covered is also narrower. I have deliberately omitted a number of complicating details in the signal processing and, indeed, important statistical questions still remain unanswered regarding the best treatment of this kind of data. The essential point is that the number of available observations is rather small.

2. The Usual Inverse Problem. We consider at first the simplest system, an electrically conducting layer with conductivity $\sigma \in C^2[0,h]$; also note $\sigma > 0$. Electromagnetic fields are excited in the medium by a periodic source varying as $e^{i\omega t}$. In the medium the complex electric field E obeys the ordinary differential equation

$$\frac{d^2E}{dz^2} = i\omega\mu_0\sigma E \tag{1}$$

where μ_0 is a fixed constant of the MKS system of units. At $h = 0$, the bottom of the layer, the boundary condition $E = 0$ is applied, which corresponds to the proper condition at the center of a spherical Earth or at the interface with a region of infinite conductivity. Also we specify that $dE/dz\,|_h \neq 0$. If σ is known we can easily calculate the complex electromagnetic response c for any frequency ω; it is defined by

$$c = -E(0)/\left.\frac{dE}{dz}\right|_0 . \tag{2}$$

Because $dE/dz\,|_0 = i\omega B$, where B is the magnetic field at the surface, c is a quantity that can be estimated in a magnetotelluric experiment; traditionally, values are reported as apparent resistivity ρ_a and phase Φ, where

$$c = \left(\frac{\rho_a}{\mu_0\omega}\right)^{1/2} e^{i(\Phi-\pi/2)}$$

Considered as a function of real frequency, c is a complex function, analytic at all points except possibly $\omega = 0$. The usual inverse problem is of course to discover what can be said about σ when the function c is known.

Weidelt [1972] gave the necessary transformations to reduce the problem to a conventional Schrödinger-style inverse scattering problem which he then solved with the Gel'fand-Levitan machinery (strictly Weidelt-treated $\sigma \in C^2(0,\infty)$). To make the connection with the spectral inverse problem a little clearer let us perform some simple calculations. Consider the Green's function $G(z,z_0)$ for (1) under the boundary conditions $dE/dz|_0 = E(h) = 0$:

$$\frac{d^2G}{dz^2} - i\omega\mu_0\sigma G = \delta(z-z_0) .$$

It is easily shown that

$$c = -G(0,0) .$$

Under the assumptions about σ, it is elementary that (1) possesses an infinite number of real eigensolutions u_n, complete in $L_2(0,h)$ satisfying the boundary conditions $u_n(0) = 0$, $du_n/dz|_h = 0$ and

$$\int_0^h u_n^2 \, dz = 1$$

$$\frac{du_n}{dz^2} + \lambda_n u_n = 0 , \quad n = 1, 2, 3, \cdots \tag{3}$$

where λ_n is real and positive. If the Green's function G is expanded in the eigenfunction basis, we find

$$G(z,z_0) = -\frac{1}{\mu_0\sigma(z_0)} \sum_{n=1}^{\infty} \frac{u_n(z)u_n(z_0)}{\lambda_n+i\omega}$$

so that

$$c = \sum_{n=1}^{\infty} \frac{[u_n(0)^2/\mu_0\sigma(0)]}{\lambda_n+i\omega}$$

$$= \sum_{n=1}^{\infty} \frac{a_n}{\lambda_n+i\omega}$$

where $a_n = u_n(0)^2/\mu_0\sigma(0) > 0$. We can express the sum as a Stieltjes integral

$$c = \int_0^{\infty} \frac{da(\lambda)}{\lambda+i\omega} \tag{4}$$

where a is bounded nondecreasing real function: it is the spectral function of the differential operator $(\mu_0\sigma)^{-1}(d/dz)^2$. An equation like (4) holds for a much more general class of conductivity functions as we shall see. Equation (4) shows that c is a linear functional of a; in fact, it also shows that a is obtainable from c by analytic continuation of c away from the real ω axis, is where it is measured, onto the positive imaginary axis. Although the original Gel'fand-Levitan treatment was for spectral inverse problems associated with a different kind of differential equation from (3), knowledge of a in (4) is equivalent to knowledge of the spectral function in the corresponding Schrödinger problem to which (1) can be transformed.

The question of conditions on c to insure existence of solutions σ can all be easily derived from those known in the classical theory. If a solution exists, it is unique. And of course there is a constructive procedure for discovering the profile σ from a given legitimate c. I shall not dwell on these matters because they represent nothing not already well known.

3. The Geophysical Problem. As discussed earlier, knowledge of c is in practice severely restricted. It is useful to consider separately two situations: Either the observations are exact but finite so that we know the N complex numbers

$$c_j = c(\omega_j)\ , \quad j = 1, 2, \ \cdots \ N\ .$$

Alternatively, we say $c_1, c_2, \ \cdots \ c_N$ are random variables with known statistics. In the second more realistic case we may assume that the c_j are independent Gaussian variables with known means and variances, although in fact the true statistical description is at present a matter of current research [e.g., Park and Chave, **1983**]. To apply the standard theory requires the "completion" of the finite data set to an analytic function of ω. Obviously this can be done in many ways, but it is difficult to see how to arrange the interpolation so as to produce a function c that will satisfy the conditions for the existence of a solution. Furthermore, the unstable character of the inverse problem, exaggerated here by the need to find a from c effectively by analytic continuation, means that the minutest details in the interpolation process may have a profound effect on the final answer, if a solution can be found at all.

The only satisfactory procedure is to begin the problem again with the finite data set in mind from the start. The first matter, which has been satisfactorily resolved, is that of existence: we must decide whether or not a given finite data set $c_1, c_2, \ \cdots \ c_N$ is compatible with the mathematical model. We ought not be too surprised if actual experimental data fail to be compatible, principally because of the simplifications of variation of σ only with depth; in a surprisingly high proportion of cases, however, the measurements prove to be consistent with the simple model. The key to the solution of this problem is (4) although it pays to introduce a generalization of the class of valid conductivity profiles [Parker, **1980**]. We allow σ to exhibit delta function behavior and we demand only that $\sigma \geqslant 0$, rather than requiring strict positivity; the simple theory outlined earlier now fails, but if we suppose

$$\tau(z) = \int_z^h \sigma(y)\,dy$$

and $\tau \in NBV$, the space of functions of bounded variation, normed by the variation, then for any τ of this class we can always find $b_0 \geqslant 0$, and b, a real, bounded nondecreasing function such that the response of the system is given by

$$c = b_0 + \int_0^\infty \frac{1-i\omega\lambda}{\lambda+i\omega}\,db(\lambda)\ . \tag{5}$$

Of course (1) must also be generalized to deal with the wider class of conductivity functions. Equation (5) is analogous to (4). Let us take first the case of exact data; since c is a *linear functional* of b_0 and b, compatibility of the system

$$c_j = b_0 + \int_0^\infty \frac{1+i\omega_j\lambda}{\lambda+i\omega_j}\,db(\lambda) \tag{6}$$

is equivalent to the feasibility of a semi-infinite linear program, because of the constraint that b must not decrease. We can construct a certain sequence of finite linear programs of increasing size, that is guaranteed to settle the question. Satisfying (6) is a necessary condition, but is it sufficient? In the sense that σ is allowed in the wider class of conductivities, yes. If any function b can satisfy (6), there is one that increases only at a finite number of points. This corresponds to a spectrum with only a finite number of eigenvalues and the appropriate model then consists of a finite sequence of delta functions in σ:

$$\sigma(z) = \sum_{j=1}^{J} \tau_j \delta(z-z_j)\ . \tag{7}$$

The problem is almost identical to the one treated by Krein **[1952]** of the free oscillations of a light string with unknown masses placed in unknown positions on it.

If we introduce noisy data the new question is whether or not there are $b_0 \geqslant 0$ and nondecreasing b such that

$$\sum_j (c_j - L_j[b_0,b])^2/\epsilon_j^2 < \chi^2 \tag{8}$$

where $L_j[b_0,b]$ is an abbreviation for the right side of (6), ϵ_j is the statistical error associated with the jth observation of c_j, and χ^2 is the smallest unacceptable level of misfit, determined according to some statistical preference. This is a problem in quadratic programming that can be solved in a similar way by discovering the smallest value of the two-norm misfit.

The conditions for the existence of a solution turn out to be extremely delicate for exact data. This is illustrated by a numerical example in which I generated the responses c_j at 15 frequencies for a simple model profile; the model and the frequencies were typical of those encountered in crustal studies (depths to 30 km). When random perturbations with amplitudes of 1 part in 10^4 were added to these artificial responses, it was found that (6) could no longer be exactly satisfied. Clearly empirical data, whose accuracy is rarely better than 1 part in 10^2, cannot be expected to be capable of satisfying (6); this confirms the assertion that direct application of the standard analytical inverse process is unlikely to be successful with actual measurements.

The matter of uniqueness of solutions is almost trivial. If (8) is accepted as a suitable criterion with noisy data, there can be no question about it — if model exists obeying (8), then infinitely many others can. With exact data the normal state of affairs is the same. However, I have shown that there are certain exceptional finite data sets that can support only one conductivity profile [Parker, **1980**].

Most research into the magnetotelluric inverse problem concerns the construction of profiles capable of accounting for the observations. The fact that a great variety of models may all be equally good at explaining the observations does not entirely invalidate the geophysicist's desire to have a look at one. If we use the solution found by minimizing the two-norm in (8) we generate the pathological insulator and delta-function profiles of (7). The recovery of the parameters of these systems from the spectral function is an interesting problem in numerical analysis — it is the question of constructing the coefficients of a continued fraction from a given equivalent partial fraction representation; interested readers are referred to Parker and Whaler [**1981**]. These best-fitting models are quite unreasonable physically. We should be willing to accept a slightly worse fit to the measurements in exchange for a smoother model. The selection of a particular profile from the manifold of acceptables ones [i.e., those obeying (8)] invites us to exercise our geophysical or geological prejudices about the kind of model to be expected. In shallow regions it is observed that abrupt changes of factors of 100 or more in conductivity occur at interfaces between different rock types or at the edges of water-saturated zones. Thus one may plausibly argue that models relying on simple discontinuities for their variation might be most suitable. A common parameterization is to allow only a small number of homogeneous layers [e.g., Fischer and LeQuang, **1981**]. In my own work on this problem I have taken the view that it is important to be sure that the construction process is capable of yielding solutions whose misfit to the observations can be made as close to the minimum value as desired. The best-fitting solution is an end member of a single-parameter family of models: when the parameter is small, we approach the best-fitting but highly variable solutions; at the other extreme the models are quite featureless, but (usually) they fail to satisfy the observations. We define the class of piecewise constant functions:

$$\sigma(z) = \sigma_n,\ z_n \leqslant z < z_{n+1}$$

where

$$z_{n+1} - z_n = h_n = d/\sqrt{\sigma_n \mu_0}$$

and

$$z_1 = 0\ .$$

Thus $\sigma_n h_n^2 = \mu_0 d^2$ in every layer of this stack of uniform conductors, and d is the free parameter. For a fixed parameter d, the best fitting solution in this class can be constructed using techniques that are almost the same as those used for building delta-function models because c can be expressed as a rational function of $\cosh d(i\omega)^{1/2}$. When d is very small, the computed models consist of alternating thin, highly conducting layers and wide resistive ones.

Deeper within the Earth it is believed the conductivity is controlled by temperature and that smoothly varying solutions are more likely than discontinuous ones. I have developed another construction scheme for generating such models involving a single free parameter and based upon the Gel'fand-Levitan solution. It is one way to treat the idea of interpolation of the observations in a systematic manner. We write

$$c(\omega) = \frac{1}{\sqrt{i\omega\mu_0\sigma_0}} + \int_0^\infty \frac{d\bar{a}(\lambda)}{\lambda + i\omega}$$

where σ_0 is the parameter which we may choose and $\bar{a}$ is a nondecreasing real bounded function. We optimize the fit of this functional form to the observations in the least-squares sense by choosing the appropriate $\bar{a}$ by a quadratic program like that used to satisfy (8). The resulting spectral function leads to a degenerate kernel in the Fredholm equation of Gel'fand and Levitan, thus permitting efficient stable computations. In this case, large σ_0 leads to models that fit the data very well; they then consist of a series of tall thin peaks in the conductivity.

Other construction algorithms for generating the very smooth profiles are under development, but we shall not spend any time on them. I want to turn to what I consider to be the most important question, but it is the one on which the least progress has been made.

4. Inference. We must accept the nonuniqueness of the solution, but this does not mean that the observations are devoid of information about the Earth; the fundamental problem is to discover just what information is available. In linear inverse problems (for example tomography) when the number of observational constraints is finite and the models reside in an infinite-dimensional linear vector space, the traditional way of assessing the reliability of a model or, equivalently, of making an error estimate, is to introduce a norm on the model space and to assume that an upper bound for the norm is known. A difficulty here is that it is very hard to give a useful bound for the conductivity. Near the surface, where drilling can sample the materials, we find four or five orders of magnitude variations in conductivity of materials, because water is a good conductor and most minerals are excellent insulators. At greater depths we have less information, and the danger of circularity arises — the bounds may be derived from the inversion of electromagnetic data.

The safest way to proceed in my opinion is to find out what can be inferred without introducing additional limitations from outside the set of observations. For linear inverse problems this may be impossible — confining solutions to a ball in model space is just a way of bringing a *nonlinear* constraint into the linear inverse problem. The magnetotelluric inverse problem is already nonlinear. There is a definite limit on what can be learned from the data alone [Parker, **1982**]. If a perfectly conducting layer ($\sigma = \infty$) is introduced at a certain depth, whatever is below that level has no electrical influence at the surface. Thus, if a solution to the magnetotelluric inverse problem satisfying the observations is found that exhibits such a layer, it is logically impossible to deduce what lies under it from the response measurements. This is the basis of the argument that leads to the conclusion that there is a definite depth of total ignorance associated with any given finite data set.

Even when we investigate the region above the limiting depth, we cannot expect the observations to tell us anything useful about σ at a point. We should be able to determine bounds on the average value of σ in some interval. One way to attack the question is to treat it as a nonlinear optimization problem in which (8) appears as a constraint and the penalty function is the desired bound [Oldenburg, **1983**]. Unfortunately, no rigorous theory at present exists for this approach; it is just a massive numerical search in which, as far as we know, false local minima may exist to deceive the investigator.

BIBLIOGRAPHY

Fischer, G. and B.V. LeQuang, *Topography and minimization of the standard deviation in one-dimensional magnetotelluric modelling*, Geophys. J. Roy astr. Soc. **67** (1981), 279–292.

Krein, M.G., *On inverse problems for inhomogeneous strings*, Dokl. Akad. Nauk. SSSR **82** (1952), 669–672.

Oldenburg, D.W., *Appraisal in linear and nonlinear inverse problems*, J. Geophys. Res. in press, 1983.

Park, J. and A.D. Chave, *On the estimation of magnetotelluric response functions using singular value decomposition*, J. Geophys. Res. in press, 1983.

Parker, R.L., *The inverse problem of electromagnetic induction: Existence and construction of solutions based upon incomplete data*, J. Geophys. Res. **85** (1980), 4421–4428.

Parker, R.L., *The existence of a region inaccessible to magnetotelluric sounding*, Geophys. J. Roy astr. Soc. **68** (1982), 165–170.

Parker, R.L. and K. Whaler, *Numerical methods for establishing solutions to the inverse problem of electromagnetic induction*, J. Geophys. Res. **86** (1981), 9574–9584.

Rokityansky, I.I., *Geoelectric Investigation of the Earth's Crust and Mantle*, Springer, 1982.

Weidelt, P., *The inverse problem of geomagnetic induction*, Z. Geophys. **38** (1972), 257–289.

INSTITUTE OF GEOPHYSICS AND PLANETARY PHYSICS, SCRIPPS INSTITUTION OF OCEANOGRAPHY, UNIVERSITY OF CALIFORNIA, SAN DIEGO, LA JOLLA, CA 92093

SIAM-AMS PROCEEDINGS
Volume **14**
1984

APPLICATION OF THE TRACE FORMULA METHODS TO INVERSE SCATTERING FOR SOME GEOPHYSICAL PROBLEMS

D.C. Stickler

ABSTRACT. Suppose for the scalar Helmhotz wave equation for a stratified medium, the propagation speed $c(z)$ is known for $z < 0$ but unknown for $z > 0$. Trace formula methods are used to recover $c(z)$ for $z > 0$, where $c(z) \to c_1$ as $z \to -\infty$ and $c(z) \to c_2 \geqslant c_1$ as $z \to +\infty$. The speed c_2 need not be known. A point harmonic source which excites no proper modes is located in the region $z < 0$, and the field is measured for all $r > 0$ at some fixed $z_R < 0$. A reflection coefficient is recovered from this measurement and is used in the trace method to recover $c(z)$. Two numerical examples illustrate the applicability of the trace method and demonstrate that when the accuracy to which the reflection coefficient is known increases, so does the depth, z, to which $c(z)$ can be recovered.

Introduction

In 1979, P.A. Deift and E.B. Trubowitz [1] presented a new technique for solving the inverse scattering problem of quantum mechanics (i.e. the recovery of the potential $q(z)$, $-\infty < z < +\infty$, from a knowledge of the reflection coefficient $R(k)$). They called their technique the trace method, and it involves solving a non-linear ordinary differential equation, rather than the linear integral equation associated with the Marchenko method [2]. This paper is concerned with an extension of their method to a more general class of potentials, the numerical implementation of the trace method for these potentials and its application to some problems of geophysical interest.

First, a brief comparison of the trace method and the more classical Marchenko method is in order. The data for both methods is identical, namely, the reflection coefficient $R(k)$ where k is related to the quantum mechanical energy but in the geophysical problems to be discussed below to a vertical wave number. The kernel of the linear integral equation in the Marchenko method involves a Fourier transform of this reflection coefficient while in the trace

*This work was supported by the Office of Naval Research.

method the potential and reflection coefficient are related by

$$q(z) = \frac{2i}{\pi} \int_{k=-\infty}^{+\infty} k\ R(k)\ u^2(z,k)dk \ , \tag{1}$$

which is called the trace formula, where u(z,k) satisfies

$$\frac{d^2u}{dz^2} + (k^2 - q(z))u = 0 \tag{2a}$$

with

$$u(z,k) \sim e^{-ikz} \ , \ z \to -\infty \ . \tag{2b}$$

Hence the Fourier transform of the data is not required. In the Marchenko method, differentiation of the solution to the linear integral equation is required to recover the potential. For the trace method, the solution of Eq. 2a subject to Eq. 2b with q(z) in Eq. 2a replaced by Eq. 1 is required. When Eq. 2a is satisfied at some z, then Eq. 1 is used to recover q(z). This constitutes the principal differences between the two schemes. The convergence of an interation procedure to determine u(z,k) and hence q(z) and the characterization of R(k) is discussed in Reference 1 for the case when q(z) is sufficiently smooth and vanishes at infinity sufficiently rapidly.

In some geophysical applications it cannot be assumed that q(z) approaches the same known values as $|z| \to \infty$, and one purpose of this paper is to show that the trace formula in Eq. 1 can be derived for the case when q(z) approaches zero as $z \to -\infty$ and $\alpha^2 > 0$ where α^2 is unknown as $z \to +\infty$. The reflection coefficient needed in Eq. 1 is recovered from a measurement of the field for some fixed z_R and all r. The source is a point harmonic source which excites no proper modes. Finally numerical examples are presented to demonstrate the application of the method to some cases of geophysical interest.

The trace method has been applied by Greene [3] to the one-dimensional acoustic wave equation. It has also been applied by D.C. Stickler and P.A. Deift [4] to recover the sound speed profile in a half space bounded by a pressure release surface. The source in this case is a point harmonic source located below the pressure release surface and the normal derivative of the pressure field is measured at the pressure release surface. Stickler has applied the method to be discussed in the paper to the acoustic case [5] where it has been shown how to recover both the sound speed and density profiles and to the electromagnetic case [6] where it has been shown how to recover the permittivity and permeability. Coen [7] has examined similar problems using the more classical methods of Marchenko and Gel'fand-Levitan.

Three recent papers discuss this class of inverse problems from the time domain approach. These papers are by R. Burridge [8], K. Bube and R. Burridge [9] and W. Symes [10].

Trace methods have been used by C. Tomei [11] to express the two coefficients in a third order differential equation associated with the Boussinesq equation, using certain scattering data.

Statement of Problem

Let the field U(r,z) satisfy the scalar Helmholtz wave equation

$$\Delta U(r,z) + (\omega^2/c^2(z))\, U(r,z) = -\frac{\delta(r)}{2\pi r}\,\delta(z-z_s), \tag{3}$$

subject to an outgoing condition at infinity. Furthermore, assume that $c(z)$ is known for $z < 0$ but unknown for $z > 0$, and that $z_s < 0$, i.e the source is in the region where the $c(z)$ is known. Assume that at the angular frequency ω, no square integrable eigenmodes are excited. Assume that

$$c(z) \to c_1 \quad \text{as} \quad z \to -\infty \tag{4a}$$

$$c(z) \to c_2 \geq c_1 \quad \text{as} \quad z \to +\infty \tag{4b}$$

The condition in Eq. 4b does not require that c_2 be known, only that it is not less than c_1. Define $q(z)$ as follows:

$$q(z) = \frac{\omega^2}{c_1^2}\left(1-(c_1/c(z))^2\right), \tag{5}$$

and require

$$\int_{-\infty}^{z} |q(z)|dz, \qquad \int_{-\infty}^{z} |q'(z)|dz < \infty, \tag{6a}$$

and

$$\int_z^\infty |q_\alpha(z)|dz < \infty \ , \quad \int_z^\infty |q'_\alpha(z)|dz < \infty \ , \tag{6b}$$

where

$$q_\alpha(z) = q(z) - \alpha^2 \ , \tag{6c}$$

$$\alpha^2 = \frac{\omega^2}{c_1^2}\left(1-(c_1/c_2)^2\right) . \tag{6d}$$

Define $\hat{u}(\beta,z)$ to be the Hankel transform of $U(r,z)$,

$$\hat{u}(\beta,z) = \int_{r=0}^\infty U(r,z)\, J_0(\beta r) r\, dr, \tag{7a}$$

and

$$U(r,z) = \int_{\beta=0}^\infty \hat{u}(\beta,z)\, J_0(\beta r)\, \beta d\beta \ . \tag{7b}$$

Substitution of Eq. 7b into Eq. 3 shows that $\hat{u}(\beta,z)$ satisfies

$$\frac{d^2\hat{u}(\beta,z)}{dz^2} + \left(\frac{\omega^2}{c^2(z)} - \beta^2\right)\hat{u}(\beta,z) = -\frac{\delta(z-z_s)}{2\pi} \tag{8a}$$

subject to an outgoing condition as $|z| \to \infty$. Denote the solutions of the homogeneous form of Eq. (8a) by $u(z,k)$ where

$$k = \left((\omega/c_1)^2 - \beta^2\right)^{1/2}$$

i.e. $u(z,k)$ satisfies Eq. 2a, with $q(z)$ defined in Eq. 5 and k in Eq. 9a. Note that k is the vertical wave number and β the radial wave number.

Next, a measurement of the field $U(r,z)$ is related to a reflection coefficient $R(k)$. At some $z_R < 0$, the field $U(r,z)$ is measured for all $r > 0$ and Eq. 7a is used to calculate $\hat{u}(\beta,z)$ for all real β, and (it however, cannot be used for complex β), as to be described below, the reflection coefficient

$R(k)$ is determined by $\hat{u}(\beta,z)$. However, the trace formula in Eq. (1) requires that $R(k)$ be known for all real k. Eq. 9a makes it possible to relate k and β where

$$0 \leqslant \beta \leqslant \omega/c_1 \ , \tag{10}$$

and this range is the most significant physically because it corresponds to the real angles of incidence if the correspondence

$$\beta = \frac{\omega}{c_1} \sin\theta \tag{11}$$

is made. For $\beta > \omega/c_1$ this yields values of $\hat{u}(\beta,z)$ on the positive imaginary k axis. A theorem of Van Winter [12] discussed in Ref. [4] or a numerical procedure discussed in [5] can be used to continue this data to the real k-axis. The numerical examples considered below show that the range of β in Eq. 10, i.e. the real angles, are adequate for some purposes.

The relationship between $\hat{u}(\beta,z_R)$ and the reflection coefficient $R(k)$ is described in terms of three independent solutions of Eq. 2a. Let $u_j(zk)$ satisfy Eq. 2a and

$$u_1(z,k) \sim e^{ik_\alpha z} \ , \quad z \to \infty \tag{12a}$$

$$u_2(z,k) \sim e^{-ikz} \ , \quad z \to -\infty \tag{12b}$$

$$u_3(z,k) = u_2(z,-k) \tag{12c}$$

where $$k_\alpha = (k^2-\alpha^2)^{1/2} \quad \mathrm{Im}k_\alpha > 0 \ . \tag{12d}$$

The branch cut for k_α extends from $\pm\ \alpha$ along the real axis to $\pm\ \infty$, respectively, (α is non-negative since $c_2 \geqslant c_1$). Note that $u_2(z,k)$ is used in Eqs. 1,2a and 2b to determine $q(z)$.

The Wronskian of $u_2(z,k)$ and $u_2(z,-k)$ is given by

$$W(u_2(z,k),\ u_2(z,-k)) = 2ik \ . \tag{13}$$

The three solutions are related by

$$u_2(z,-k) + R(k)u_2(z,k) = T(k)u_1(z,k) \tag{14}$$

where the constants $R(k)$ and $T(k)$ are identified as reflection and transmission coefficients because of the asymptotic behavior of $u_1(z,k)$, $u_2(z,k)$ and $u_2(z,-k)$. From Eqs. 13 and 14, the Wronskian of $u_1(z,k)$ and $u_2(z,k)$ is given by

$$W(u_1(z,k)u_2(z,k)) = -2ik/T(k) \tag{15}$$

and therefore

$$\begin{aligned} \hat{u}(\beta,z_R) &= \frac{T(k)}{2ik}\,\frac{u_2(z,k)u_1(z_s,k)}{2\pi}, \quad z < z_s \\ &= \frac{T(k)}{2ik}\,\frac{u_2(z_s,k)u_1(z,k)}{2\pi}, \quad z > z_s \,. \end{aligned} \tag{16}$$

Suppose, $z_R < z_s$ then substitution from Eq. (14) for $T(k)u_1(z_s,k)$ into Eq. 16 yields

$$\hat{u}(\beta,z_R) = \frac{1}{(2\pi)(2ik)}\, u_2(z_R,k)\left[u_2(z_s,-k) + R(k)\, u_2(z_s,k)\right] \tag{17}$$

In Eq. 17, $\hat{u}(\beta,z_R)$ is known from the measurement, $u_2(z_R,k)$, $u_2(z_s,-k)$ and $u_2(z_s,k)$ can be calculated since $u_2(z,\pm k)$ satisfy initial conditions at $-\infty$ and $c(z)$ is known for $z < 0$. Thus, $R(k)$ is determined in terms of the measured data by Eq. 17.

Trace formula Derivation

The trace formula, Eq. 1, is derived using a modification of the method used in Ref. 1, and in order to carry out this derivation, expansions of $u_1(z,k)$, $u_2(z,k)$ and $T(k)$ must be obtained which are valid as $|k| \to \infty$, $\operatorname{Im}k \geq 0$. The method used in Ref. 1 fails because $q(z)$ approaches α^2 rather than zero as

$z \to \infty$. These expansions are obtained using a modification of the usual Jost function approach .

The expansion for $u_2(z,k)$ is derived first. Let $u_2(z,k)$ be written

$$u_2(z,k) = m_2(z,k)\exp\{-ikz - \frac{1}{2ik}\int_{z'=-\infty}^{z} q(z')dz'\}, \tag{18}$$

where $m_2(z,k)$ satisfies the initial conditions (use Eqs. 12b and 5 and the fact that $q(z) \to 0$ as $z \to -\infty$)

$$m_2(z,k) \sim 1, \quad z \to -\infty \;, \tag{19a}$$

$$m_2'(z,k) \sim 0, \quad z \to -\infty \;. \tag{19b}$$

Substitution of Eq. 18 in Eq. 2a shows that $m_2(z,k)$ satisfies

$$m_2''(z,k) - 2(ik + \frac{q}{2ik})m_2' + (\frac{-q'}{2ik} + \frac{q^2}{(2ik)^2})m_2 = 0, \tag{19c}$$

therefore, $m_2(z,k)$ satisfies the following Volterra integral equation

$$m_2(x,k) = 1 + \int_{z'=-\infty}^{z} K_2(z,z')\{\frac{q'(z')}{2ik} - \frac{q^2(z')}{2ik)^2}\}m_2(z',k)dz', \tag{19d}$$

where

$$K_2(z,z') = \int_{z''=z'}^{z} \exp(2\{ik(z''-z') + \frac{1}{2ik}\int_{z'''=z'}^{z''} q(z''')dz'''\})dz''. \tag{19e}$$

Note that as $|k| \to \infty$, $\mathrm{Im}k \geqslant 0$, the kernel is $O(k^{-1})$ since $z'' > z'$. Successive iterations of Eq. 19d converge uniformly to $m_2(z,k)$. Integration by parts shows that $K_2(z,z')$ can be expanded as

$$K_2(z,z') = \exp\left(\frac{1}{2ik}\int_{z''=z'}^{z} q(z'')dz''\right) \cdot$$

$$\cdot \frac{e^{2ik(z-z')}}{2ik} - \frac{1}{2ik} + O(k^{-3}) \quad , \ \text{Im}k \geqslant 0, \ |k| \to \infty \tag{20a}$$

One iteration of Eq. (19d) and the use of the Riemann-Lebesgue theorem shows that

$$m_2(z,k) = 1 - \frac{q(z)}{(2ik)^2} + O(k^{-2-v}) \ , \ v > 0. \tag{20b}$$

The expansion in Eq. 20a has been used. A similar expansion is obtained for $m_1(z,k)$ below, and v is then the smaller of the two decay rates. Substitution of Eq. 20b into Eq. 19c shows that Eq. 19c is satisfied to $O(k^{-1-v})$, Im$k \geqslant 0$, $|k| \to \infty$. The initial conditions in Eq. 19a and b are also satisfied.

In a similar manner $m_1(z,k)$ is defined by

$$u_1(z,k) = m_1(z,k) \exp\left(ik_\alpha z - \frac{1}{2ik_\alpha}\int_{z'=z}^{\infty} q_\alpha(z')dz'\right) \tag{21a}$$

where

$$m_1''(z,k) + 2(ik_\alpha + \frac{q_\alpha}{2ik_\alpha})\, m_1'(z,k) + \left(\frac{q_\alpha'(z')}{2ik_\alpha} + \frac{q_\alpha^2}{(2ik_\alpha)^2}\right) m_1(z,k) = 0 \ , \tag{21b}$$

where

$$m_1(x,k) \to 1, \text{ and } m_1'(z,k) \to 0, \ z \to \infty \ . \tag{21c}$$

The Volterra integral equation satisfied by $m_1(z,k)$ is given by

$$m_1(z,k) = 1 - \int_{z'=z}^{\infty} K_1(z,z') \left\{\frac{q_\alpha'(z')}{2ik_\alpha} + \frac{q_\alpha^2(z')}{(2ik_\alpha)^2}\right\} m_1(z',k)dz' \ , \tag{22a}$$

where

$$K_1(z,z') = \int_{z''=z}^{z'} \exp\left(2\left\{ ik_\alpha(z'-z'') - \frac{1}{2ik_\alpha}\int_{z'''=z''}^{z'} q_\alpha(z''')dz'''\right\}\right)dz'' . \qquad (22b)$$

Again, sucessive iterations converge uniformity to $m_1(z,k)$ for $\mathrm{Im}k \geq 0$. The kernel $K_1(z,z')$ has the expansion

$$K_1(z,z') = \frac{-1}{2ik_\alpha} + \frac{1}{2ik_\alpha} e^{2ik_\alpha(z'-z)}\left(-\frac{1}{2ik_\alpha}\int_{z'''=z}^{z'} q_\alpha(z''')dz'''\right) +$$

$$O(k_\alpha^{-3}) , \ \mathrm{Im}k \geq 0, \ |k| \to \infty , \qquad (23a)$$

and one iteration of Eq. 22a (using $K_1(z,z')$ from Eq. 23a) shows that

$$m_1(z,k) = 1 - \frac{q_\alpha(z)}{(2ik_\alpha)^2} + O(k_\alpha^{-2-v}), \ v > 0, \ \mathrm{Im}k \geq 0, \ |k| \to \infty . \qquad (23b)$$

Substitution of Eq. 23b into 21b shows that Eq. 21b is satisfied to $O(k^{-1-v})$. The initial conditions in Eq. 21c are also satisfied.

Next an expansion of $T(k)$ is obtained for $\mathrm{Im}k \geq 0$, $|k| \to \infty$ by observing from Eq. 14 that the coefficient of $\exp(ikz)$ as $z \to -\infty$ of $u_1(z,k)$ is $T^{-1}(k)$. Fix z_0, then the expression for $u_1(z,k)$ (substitution of Eq. 23b into 21a) can be written

$$u_1(z,k) = \left(1 - \frac{q_\alpha(z)}{(2ik_\alpha)^2}\right) \exp\left\{ ik_\alpha z - \frac{1}{2ik_\alpha}\int_{z'=z}^{z_0} q_\alpha(z')dz' - \frac{1}{2ik_\alpha}\int_{z'=z_0}^{\infty} q_\alpha(z')dz'\right\}$$

$$+ O(k^{-2-v})$$

$$= \left(1 - \frac{q_\alpha(z)}{(2ik_\alpha)^2}\right) \exp\left\{\left(ik_\alpha - \frac{\alpha^2}{2ik_\alpha}\right)z + \frac{\alpha^2}{2ik_\alpha} z_0 - \frac{1}{2ik_\alpha}\int_{z'=z}^{z_0} q(z')dz'\right.$$

$$\left. - \frac{1}{2ik_\alpha}\int_{z'=z_0}^{\infty} q_\alpha(z')dz'\right\} .$$

$$+ O(k^{-2-v}), \ \mathrm{Im}k \geq 0 .$$

However,

$$ik_\alpha - \frac{\alpha^2}{2ik_\alpha} = ik + O(k_\alpha^{-3}) \tag{24}$$

and therefore

$$u_1(z,k) = \left(1 - \frac{q_\alpha(z)}{(2ik_\alpha)^2}\right) \exp\left\{ikz + \frac{\alpha^2 z_0}{2ik_\alpha} - \frac{1}{2ik_\alpha}\int_{z'=z}^{z_0} q(z')dz'\right.$$

$$\left. - \frac{1}{2ik_\alpha}\int_{z'=z_0}^{\infty} q_\alpha(z')\right\} + O(k^{-2-v}),\ \mathrm{Im}k \geqslant 0\ .$$

From the observation above

$$T^{-1}(k) = \left(1 + \frac{\alpha^2}{(2ik_\alpha)^2}\right) \exp\left\{\frac{\alpha^2 z_0}{2ik_\alpha} - \frac{1}{2ik_\alpha}\int_{z'=-\infty}^{z_0} q(z')dz'\right.$$

$$\left. - \frac{1}{2ik_\alpha}\int_{z'=z_0}^{\infty} q_\alpha(z')\right\} + O(k^{-2-v}),\ \mathrm{Im}k \geqslant 0\ ,$$

and

$$T(k) = \left(1 - \frac{\alpha^2}{(2ik_\alpha)^2}\right) \exp\left\{-\frac{\alpha^2 z_0}{2ik_\alpha} + \frac{1}{2ik_\alpha}\int_{z'=-\infty}^{z_0} q(z')dz' + \right.$$

$$\left. \frac{1}{2ik_\alpha}\int_{z'=z_0}^{\infty} q_\alpha(z')\right\} + O(k^{-2-v}),\ \mathrm{Im}k \geqslant 0\ . \tag{25}$$

The product $T(k)u_1(z,k)u_2(z,k)$ is given by (Eqs. 18,20b,21a,23b,25)

$T(k)u_1(z,k)u_2(z,k) =$

$$\left(1 - q(z)\left(\frac{1}{(2ik_\alpha)^2} + \frac{1}{(2ik)^2}\right)\right) \exp\left\{\left(ik_\alpha - \frac{\alpha^2}{2ik_\alpha}\right)z - ikz\right.$$

$$\left. + \left(\frac{1}{2ik_\alpha} - \frac{1}{zik}\right)\int_{z'=-\infty}^{z} q(z')dz'\right\} + O(k^{-2-v}),\ \mathrm{Im}k \geqslant 0\ .$$

From Eq. 24 and the observation

$$\frac{1}{k_\alpha} = \frac{1}{k} + O(k^{-3}) \tag{26}$$

the product becomes

$$T(k)u_1(z,k)u_2(z,k) = 1 - \frac{q(z)}{2(ik)^2} + O(k^{-2-v}) \ , \ \mathrm{Im}k \geqslant 0. \tag{27}$$

The integration

$$\int_{C_a} k(T(k)u_1(z,k)u_2(z,k)-1) = \frac{-i\pi}{2} q(z) + O(a^{-v})$$

where C_a is a semicircular contour of radius a integrated in the upper half k-plane clockwise from π to 0. The assumptions made earlier ($c_2 \geqslant c_1$ which implies that the k_α branch lies on the real axis, and the that no proper eigenfunctions are present) means that the integration can be performed on the real k axis between -a and +a , i.e.

$$q(z) = \frac{2i}{\pi} \int_{k=-a}^{+a} k(T(k)u_1(z,k)u_2(z,k)-1)dk + O(a^{-v}).$$

Substitution for $T(k)u_1(z,k)$ from Eq. 14 yields

$$q(z) = \frac{2i}{\pi} \int_{k=-a}^{+a} k(R(k)u_2^2(z,k) + u_2(z,k)u_2(z,-k)-1)dk + O(a^{-v}) .$$

The odd terms, i.e. $k\{u_2(z,k)u_2(z,-k)-1\}$, integrate to zero, and letting $a \to \infty$ yields the result in Eq. 1.

There is one final remark before discussing the numerical implementation of trace methods. For the initial conditions in Eq. 2b one component of the solution grows exponentially when $k^2 < q(z)$. The properties of the reflection coefficient however cancel this component. This is demonstrated in an example in ref [5]. However, if the accuracy of the reflection coefficient is fixed, then for some $z > 0$ the exponentially growing factor causes the calculations to become inaccurate for z sufficiently large.

Numerical Method

Rewrite Eq. 2a as follows

$$u''(z,k) + (k^2 - Q(u))u(z,k) = 0, \tag{28a}$$

$$Q(u) = \frac{2i}{\pi} \int_{k=-\infty}^{+\infty} kR(k)u^2(z,k)dk \ , \tag{28b}$$

subject to the initial condition

$$u \sim e^{-ikz}, \ z \to -\infty \ . \tag{28c}$$

Eqs. 28a,b,c can be written as the following Volterra integral equation

$$u(z,k) = u(z_0,k) \cos k(z-z_0) + u'(z_0,k) \frac{\sin k(z-z_0)}{k}$$
$$+ \int_{z'=z_0}^{z} \frac{\sin k(z-z')}{k} Q(u)u(z',k)dz' \ . \tag{29}$$

Let $z = z_0+h$, $u^+ \equiv u(z_0+h,k)$, $u^- = u(z_0,k)$, $u^{+'} \equiv \frac{du}{dz}(z_0+h,k)$, $u^{-'} = \frac{d}{dz} u(z_0,k)$, $Q^+ = Q(u^+)$, $Q^- = Q(u^-)$, then Eq. 29 and its derivative with respect to z can be written

$$\begin{bmatrix} 1 & 0 \\ -\frac{h^2}{2} Q^+ & 1 \end{bmatrix} \begin{bmatrix} u(z+h) \\ hu'(z+h) \end{bmatrix} =$$

$$\begin{bmatrix} \cos(kh)+\frac{h^2}{2} Q^- \frac{\sin(kh)}{kh} & \frac{\sin(kh)}{kh} \\ -kh\ \sin(kh)+\frac{h^2}{2} Q^- \cos\ (kh) & \cos(kh) \end{bmatrix} \begin{bmatrix} u(z) \\ hu'(z) \end{bmatrix} + O(h^3) \tag{31a}$$

$$\begin{bmatrix} u(z+h) \\ hu'(z+h) \end{bmatrix} = M \begin{bmatrix} u(z) \\ hu'(z+h) \end{bmatrix} \tag{31b}$$

$$\begin{aligned} M_{11} &= \cos\ (kh) + \frac{h^2}{2} Q^- \frac{\sin(kh)}{kh} = O(h^3) \\ M_{12} &= \frac{\sin\ (kh)}{kh} = O(h^3) \\ M_{21} &= \left((kh)^2 - \frac{h^4}{4} Q^+Q^-\right) \frac{\sin\ (kh)}{kh} = O(h^3) \\ M_{22} &= \cos\ (kh) + \frac{h^2}{2} Q^+ \frac{\sin\ (kh)}{kh} = O(h^3) \end{aligned} \tag{31c}$$

The eigenvalues of M are

$$\lambda = \beta \pm (\beta^2-1)^{1/2} \tag{31a}$$

where

$$\beta = \cos kh + \frac{h^2}{2} \frac{Q^++Q^-}{2} \frac{\sinkh}{kh}, \tag{31b}$$

and $|\lambda| \leq 1$ if $\beta^2 \leq 1$ and one eigenvalues is larger than 1 for $\beta > 1$. For kh small this condition is

$$\frac{Q^+ + Q^-}{2} > k^2, \tag{32}$$

which is not an unexpected result, since one solution of Eq. 2a,b does grow exponentially when $k^2 < q$.

The iteration scheme is as follows: assume u^-, $u^{-\prime}$ and Q^- are known, guess u^+, calculate Q^+ and M, determine a new u^+. An initial guess for u^+ can be either e^{-ikz} or determined from an extrapolated value of q(z). In the numerical examples considered in the next section only a few iterations were required, i.e. usually one or two, for each u^+.

6. Numerical Examples

In this last Section two numerical examples are considered. In both c(z) is taken to be real. Furthermore, the integrations on k are restricted to $0 \leq k \leq \omega/c_1$, i.e. the real incidence angles.

In the first example

$$\begin{aligned} c(z) &= c_1, \quad z < 0 \\ &= c_1 + (c_2 - c_1)\left(3\left(\tfrac{z}{L}\right)^2 - 2\left(\tfrac{z}{L}\right)^3\right), \quad 0 \leq z \leq L \\ &= c_2, \quad L < z \end{aligned} \tag{33}$$

where c_1 = 5000 ft/sec, c_2 = 5300 ft/sec and L = 300 feet. A frequency of 50 Hz. is used.

In the second example

$$\begin{aligned} c(z) &= c_1, \quad z \leq 0 \\ &= c_i + (c_{i+1} - c_i)\left[3\left(\frac{z-z_i}{z_{i+1}-z_i}\right)^2 - 2\left(\frac{z-z_i}{z_{i+1}-z_i}\right)^3\right], \quad z_i < z \leq z_{i+1}, \ i=1,2,3 \\ &= c_4, \quad z_4 \leq z \end{aligned} \tag{34}$$

where $c_1 = 5000$, $c_2 = 5150$, $c_3 = 5050$, $c_4 = 5300$, $z_1 = 0$, $z_2 = .25L$, $z_3 = .5L$, $z_4 = L$, and $L = 300$, and the frequency is 50 Hz.

Several comments are in order. Neither profile can support a proper mode of any frequency. The reflection coefficient in each case was generated by solving a Ricatti equation directly for the reflection coefficient, for $0 \leqslant k \leqslant \omega/c_1$. In Table 1 a comparison is shown between the exact c(z) and that obtained by the trace formula for three different local tolerances (10^{-2}, 10^{-3} and 10^{-6}) for the integration determining the reflection coefficient . Roughly speaking the reflection coefficient is determined to 2, 3 and 6 significant figures. Figure 2 shows the comparison for the monotone profile in Eq. 33 and the results for the profile in Eq. 34 are shown in Fig. 3. Note that the more accurately R(k) is known the deeper it is possible to recover c(z).

	SOUND SPEED			
	TRACE FORMULA			
Z/L	TOL=10^{-2}	10^{-3}	10^{-6}	EXACT
0	5002.	5000.	5000.	5000.
.1	5009.	5008.	5008.	5008.
.2	5030.	5031.	5031.	5031.
.3	5062.	5065.	5065.	5065.
.4	5103.	5105.	5105.	5106.
.5	5148.	5149.	5150.	5150.
.6	5186.	5191.	5194.	5194.
.7	5208.	5226.	5233.	5235.
.8	5183.	5243.	5262.	5269
.9	5070.	5213.	5272.	5292.
1.0	4861.	5082.	5150.	5300.

TABLE 1. Numerical comparison of the reconstructed sound speed profile with the exact profile for a reflection coefficient determined with a local accuracy of 10^{-2}, 10^{-3}, 10^{-6}.

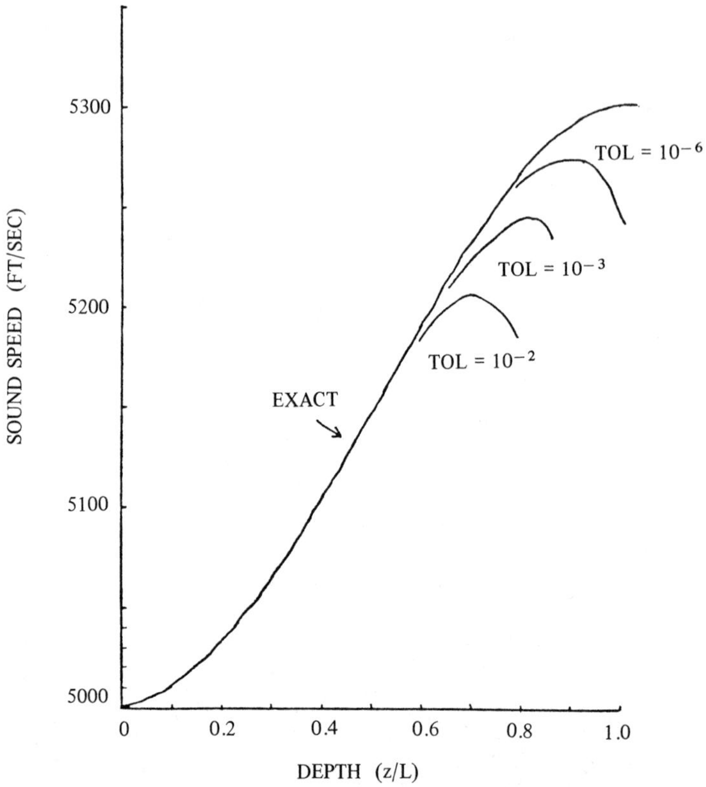

FIG. 1. A Graphical comparison of the reconstructed and exact sound speed profiles is shown only where it differs from the exact. The exact profile is described in Eq. 33.

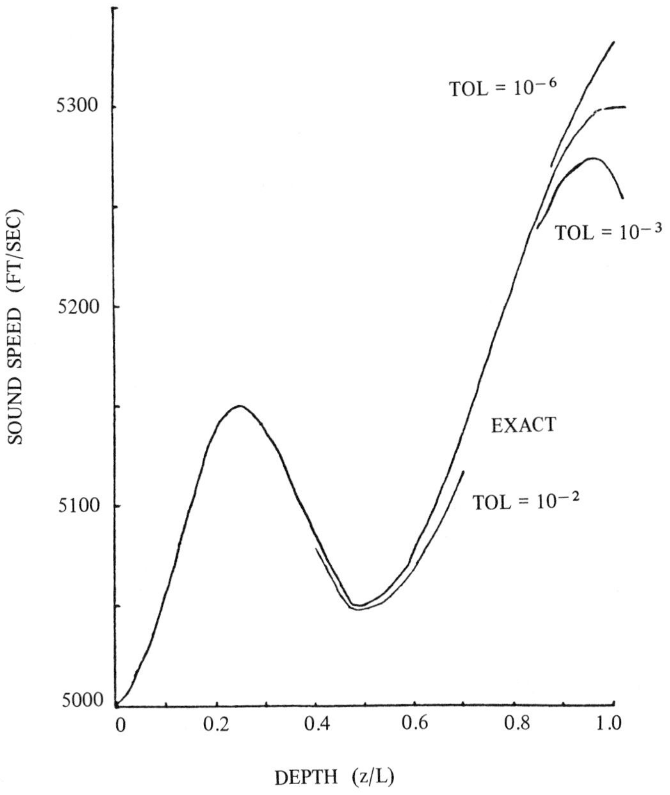

FIG. 2. A Graphical comparison of the reconstructed and exact sound speed profiles is shown only where it differs from the exact. The exact profile is described in Eq. 34.

ACKNOWLEDGEMENTS

The author would like to acknowledge the continuing encouragement of P.A. Deift and E.B. Trubowitz.

REFERENCES

[1] P. Deift, E. Trubowitz, "Inverse scattering on the line," Comm. Pure Appl. Math, Vol. XXXII (1979), p. 121-251.

[2] L.D. Faddeev, "The inverse problem in the quantum theory of scattering," J.M.P. 4, 1963, pp. 72-104.

[3] R.R. Greene, "Constructive solutions and characterization of the inverse scattering problem for the one-dimensional acoustic wave equation." Thesis. New York University (1979).

[4] D.C. Stickler and P. Deift, "Inverse problem for a stratified ocean and bottom," J. Acoust. Soc. Am. 7, (1981), p. 1723-1727.

[5] D.C. Stickler, "Inverse Acoustic Scattering for a stratified medium," Accepted Jour. Acoust. Soc. Am.

[6] D.C. Stickler, "Inverse electromagnetic scattering for a stratified medium," Inverse Optics, Proceedings of SPIE conference April 1983 Vol. 413.

[7] S. Coen, "Velocity and density profiles of a layered acoustic medium from common source point data," Geophysics, Vol. 47, 1982, pp. 898-905.

[8] R. Burridge, "The Gel′fand-Levitan, the Marchenko, and the Gopinath-Sondhi integral equations of inverse scattering theory, regarded in the context of inverse impulse response problems," Wave Motion 2 (1980), pp. 305-323.

[9] K.P. Bube and R. Burridge, "The one-dimensional inverse problem of reflection seismology," to appear SIAM review.

[10] W.W. Symes, "Impedance profile inversion via the first transport equation," J. Math. Ana. and Appl., to appear.

[11] C. Tomei, "The Boussinesq Equation", Ph.D. Thesis, N.Y.U. 1982.

[12] C. Van Winter, "Fredholm equations on a Hilbert space of analytic functions", Trans. Am. Math. Soc. 162, 103-139 (1971).

Courant Institute of Mathematical Sciences
251 Mercer Street
New York, N. Y. 10012

Part II. Computed Tomography and Inverse Problems in Medicine

SIAM-AMS PROCEEDINGS
Volume 14
1984

RADON'S PROBLEM - OLD AND NEW

A.M. Cormack

ABSTRACT: Some new results are given for Radon's problem on curves in the plane which are given, for a fixed (p,ϕ) by $r^{\alpha}\cos\{\alpha(\theta-\phi)\} = p^{\alpha}$, α real, $\alpha \neq 0$.

The first half of the talk on which this article is based was devoted to historic applications of Radon's problem in many branches of science. Since most of this material was published recently [4] it will not be repeated here. However, two recent applications, one in solid-state physics [8], and one in oceanography [1] should be noted.

Radon's problem for the families of curves

$$r^{\alpha}\cos\{\alpha(\theta-\phi)\} = p^{\alpha}, \quad \alpha > 0, \quad |\theta-\phi| < \pi/2\alpha \tag{1a}$$

and

$$p^{\beta}\cos\{\beta(\theta-\phi)\} = r^{\beta}, \quad \beta > 0, \quad |\theta-\phi| < \pi/2\beta \tag{1b}$$

have previously been studied [3] in the circular harmonic expansion

$$f(r,\theta) = \sum_{\ell=-\infty}^{+\infty} f_{\ell}(r)\, e^{i\ell\theta}, \quad \hat{f}(p,\phi) = \sum_{\ell=-\infty}^{+\infty} \hat{f}_{\ell}\,(p)\, e^{i\ell\phi}.$$

Here, $f(r,\theta)$ is a smooth, rapidly decreasing function in the plane, and $\hat{f}(p,\phi)$ is the Radon transform of f, i.e. the line integral of f along the α-curve (1a) or β-curve (1b), specified by (p,ϕ). The results for the two families of curves are quite similar because a β-curve can be obtained from an α-curve with $\alpha = \beta$ by inversion in the unit circle $((r,\theta) \to (1/r,\theta))$ and vice versa. Hence, results for the α-curves only will be discussed, and the results for the β-curves will be stated when appropriate.

1980 Mathematics Subject Classification, Primary 44A05, 45A05

On making the substitutions

$$F_\ell(s) = (1/\alpha)\ f_\ell(s^{1/\alpha})\ s^{(1/\alpha)-1} \tag{3a}$$

$$\hat{F}_\ell(q) = \hat{f}_\ell(q^{1/\alpha}) \tag{3b}$$

the equations to be solved were shown to be

$$\hat{F}_\ell(q) = 2\int_q^\infty F_\ell(s)\,\frac{\cos\{(\ell/\alpha)\cos^{-1}(q/s)\}ds}{(1-(q/s)^2)^{1/2}} \tag{4}$$

and their solutions were shown to be

$$F_\ell(t) = -\frac{1}{\pi t}\int_t^\infty \hat{F}'_\ell(q)\,\frac{\cosh\{(\ell/\alpha)\ \cosh^{-1}\ (q/t)\}da}{((q/t)^2-1)^{1/2}} \tag{5}$$

where the prime denotes a derivative. In what follows, ℓ will appear mainly through its occurrence in the cos or cosh terms, so it will not matter whether ℓ is positive or negative. Hence, to save writing, ℓ will be written instead of $|\ell|$; the context will usually make clear what is meant.

The only advantage of (5) as a solution of Radon's problem is that it explicitly demonstrates the "hole theorem", namely that in order to calculate $f(r,\phi)$ it is only necessary to know $f(p,\phi)$ for $p \geqslant r$. The grave disadvantage of (5) is that it is unstable. We can write the kernel of (5) as follows:

$$\frac{\cosh\{\gamma\cosh^{-1}x\}}{(x^2-1)^{1/2}} = \frac{\{x-(x^2-1)^{1/2}\}^\gamma + \{x + (x^2-1)^{1/2}\}^\gamma}{2(x^2-1)^{1/2}}\,. \tag{6}$$

For $x \gg 1$ this behaves like $(2x)^{\gamma-1}$, so any noise in the data $\hat{F}'_\ell(q)$ for large q is badly propagated into the calculation of $F_\ell(t)$.

In the case of the ordinary Radon transform ($\alpha = 1$) this instability is well known. Perry [9] obtained the stable solution for this case

$$F_\ell(r) = \frac{1}{\pi r}\left[\int_0^r \hat{F}'_\ell(q)U_{\ell-1}(q/r)dq - \int_r^\infty \hat{F}'_\ell(q)\frac{\{(q/r)-((q/r)^2-1)^{1/2}\}^\ell dq}{((q/r)^2-1)^{1/2}}\right] \tag{7}$$

where $U_k(x)$ is the Tschebycheff polynomial of the second kind [7]. It will be noticed that stability has been achieved by including data from the "hole", $0 \leqslant q \leqslant r$. Hansen [6] obtained (7) by using Mellin transforms, and the author (unpublished) obtained the same result using the so-called "consistency conditions" which state that $\int_0^\infty F_\ell(q)q^k dq = 0$ for certain values of k.

The purposes of this paper are first, to give the consistency conditions for the α-curves, second to show how these may be used to find stable solutions in place of (5), and third, to show that for certain values of α these stable solutions can be summed to give closed form solutions to this Radon problem.

If (4) is multiplied by q^γ and integrated from 0 to ∞, if the order of integration on the r.h.s. is interchanged, and if one uses the well-known result for the integral $\int_0^{\pi/2} \cos^x\theta\cos(y\theta)d\theta$, [7], it can be shown that

$$\int_0^\infty \hat{F}_\ell(q)q^\gamma dq = 0 \tag{8}$$

if $\gamma = (\ell/\alpha)-2, (\ell/\alpha)-4, \ldots . > -1$. These are the consistency conditions and for a given α we obtain a consistency condition only if $\ell > \alpha$. To ensure a consistency condition write $\ell/\alpha = n + 1 + \delta$ where $n = 0,1,2,\ldots$, and $0 < \delta \leqslant 1$. Enumeration of the possibilities yields the result that there are $[n/2]+1$ of them, where $[x]$ is the greatest integer not exceeding x. These can be multiplied by arbitrary real numbers, and added together, and integrated by parts to yield

$$\int_0^\infty \hat{F}'_\ell(q) \sum_{k=0}^{[n/2]} a_k\, (2q/t)^{n+\delta-2k}\, dq = 0, \tag{9}$$

where the factor $(2/t)^{n+\delta-2k}$ has been introduced for later convenience. Now (6) may be rewritten [7]

$$\frac{\cosh\{\gamma\cosh^{-1}x\}}{(x^2-1)^{1/2}} = \frac{\{x-(x^2-1)^{1/2}\}^\gamma}{2(x^2-1)^{1/2}} + (2x)^{\gamma-1}\, F\left(\frac{2-\gamma}{2}, \frac{1-\gamma}{2}, 1-\gamma, \frac{1}{x^2}\right), \tag{10}$$

where F is an ordinary hypergeometric function, so

$$\frac{\cosh\{(\ell/\alpha)\cosh^{-1}x\}}{(x^2-1)^{1/2}} = \frac{\{x-(x^2-1)^{1/2}\}^{\ell/\alpha}}{2(x^2-1)^{1/2}} + \sum_{k=0}^{\infty} b_k\, (2x)^{\ell/\alpha-1-2k} \tag{11}$$

where the b_k are known coefficients. One now adds (9) to (5) and breaks the range of the integral in (9) into two parts, and substitutes (11) for the cosh term in (5) to get

$$F_\ell(t) = \frac{1}{\pi t} \int_0^t \hat{F}_\ell'(q) \sum_{k=0}^{[n/2]} a_k (2q/t)^{n+\delta-2k}\, dq$$

$$- \frac{1}{\pi t} \int_t^\infty \hat{F}_\ell'(q) \Big[\frac{\{(q/t)-((q/t)^2-1)^{1/2}\}^{\ell/\alpha}}{((q/t)^2-1)^{1/2}} + \sum_{k=0}^{[n/2]} (2^{2k} b_k - a_k)(2q/t)^{n+\delta-2k} \Big] dq$$

$$- \frac{1}{\pi t} \int_t^\infty \hat{F}_\ell'(a) \sum_{k=[n/2]+1}^{\infty} 2^{2k} b_k (2q/t)^{n+\delta-2k}\, dq. \qquad (12)$$

If we now choose

$$a_k = 2^{2k} b_k \qquad (13)$$

the second sum (which contains the "bad part" of the kernel) vanishes, and we get the stable solution

$$F_\ell(t) = \frac{1}{\pi t} \Big[\int_0^t \hat{F}_\ell'(q) \sum_{k=0}^{[n/2]} a_k (2q/t)^{n+\delta-2k}\, dq$$

$$- \int_t^\infty \hat{F}_\ell'(q) \Big\{ \frac{\{(q/t)-((q/t)^2-1)^{1/2}\}^{\ell/\alpha}}{((q/t)^2-1)^{1/2}} + \sum_{k=[n/2]+1}^{\infty} a_k (2q/t)^{n+\delta-2k} \Big\} dq \Big], \qquad (14a)$$

where

$$a_k = \frac{(-1)^k}{k!} \frac{\Gamma(n+1+\delta-k)}{\Gamma(n+1+\delta-2k)}\,. \qquad (14b)$$

Starting from what he calls an "extended circular harmonic transform", Hansen [5] (quoted in Verly [10]) obtains a result which should be identical with (14) but which is not. The source of this discrepancy is so far unexplained. For the β-curves the stable solution is:

$$F_\ell(t) = \frac{1}{\pi t} \Big[\int_0^t \hat{F}_\ell'(q) \Big\{ \frac{\{(t/q)-((t/q)^2-1)^{1/2}\}^{\ell/\alpha}}{((t/q)^2-1)^{1/2}} + \sum_{k=[n/2]+1}^{\infty} a_k (2t/q)^{n+\delta-2k} \Big\} dq$$

$$- \int_t^\infty \hat{F}_\ell'(q) \sum_{k=0}^{[n/2]} a_k\, (2t/q)^{n+\delta-2k}\, dq \Big]. \qquad (15)$$

Henceforth we consider only the special cases α or $\beta = 1/m$, $m = 1, 2, 3 \ldots$. The cosh term in (5) becomes $T_{\ell m}(q/t)$, where $T_{\ell m}$ is the Tschebycheff function of the first kind of order ℓm [7]. Furthermore, (10) and (11) are replaced by the simpler expressions

$$\frac{T_{\ell m}(x)}{(x^2-1)^{1/2}} = \frac{\{x-(x^2-1)^{1/2}\}^{\ell m}}{(x^2-1)^{1/2}} + U_{\ell m-1}(x) \tag{16}$$

The consistency conditions can be summed and integrated as before to yield

$$\int_0^\infty \hat{F}'_\ell(q) U_{\ell m-1}(q/t)dq = 0 \tag{17}$$

and the stable solutions for the α- and β-curves are, respectively,

$$F_\ell(t) = \frac{1}{\pi t}\left[\int_0^t \hat{F}'_\ell(q)U_{\ell m-1}(q/t)dq - \int_t^\infty \hat{F}'_\ell(q)\frac{\{(q/t)-((q/t)^2-1)^{1/2}\}^{\ell m}}{((q/t)^2-1)^{1/2}}dq\right] \tag{18a}$$

and

$$F_\ell(t) = \frac{1}{\pi t}\left[\int_0^t \hat{F}'_\ell(q)\frac{\{(t/q)-((t/q)^2-1)^{1/2}\}^{\ell m}}{((t/q)^2-1)^{1/2}} - \int_t^\infty \hat{F}'_\ell(q)U_{\ell m-1}(t/q)dq\right]. \tag{18b}$$

For $\alpha = 1$, i.e. $m = 1$, (18a) is just the result (7) for the ordinary Radon transform.

Recalling the origin of F_ℓ and $\hat{F}_\ell$ in circular harmonic expansions, so that

$$\hat{F}'_\ell(q) = \frac{1}{2\pi}\int_0^{2\pi} \frac{\partial \hat{F}_\ell(q,\phi)}{\partial q} e^{-i\ell\phi}\, d\phi, \tag{19}$$

we may write

$$F(t,\theta) = \sum_{\ell=-\infty}^{+\infty} F_\ell(t)e^{i\ell\theta} = \frac{1}{2\pi^2 t}\int_0^t\int_0^{2\pi} \frac{\partial\hat{F}(q,\phi)}{\partial q}\sum_{\ell=-\infty}^{+\infty} e^{i\ell(\theta-\phi)}\, U_{|\ell m|-1}(q/t)\, d\phi dq$$

$$- \frac{1}{2\pi^2 t}\int_t^\infty\int_0^{2\pi} \frac{\partial\hat{F}(q,\phi)}{\partial q}\sum_{\ell=-\infty}^{+\infty} e^{i\ell(\theta-\phi)}\frac{\{(q/t)-((q/t)^2-1)^{1/2}\}^{|\ell m|}}{((q/t)^2-1)^{1/2}}\, d\phi dq, \tag{20}$$

after interchange of the orders of integration and summation. Now $U_\ell(\cos x) = \sin[(\ell+1)x]/\sin x$, hence the first sum in (21) is of the form $\sum_{\ell=1}^{\infty} e^{i\ell\psi}\sin(\ell m\eta)$ and the second sum is of the form $1 + 2\sum_{\ell=1}^{\infty} a^\ell \cos\ell\psi$. These sums are readily evaluated and for each range of q give the same value, namely

$$\frac{U_{m-1}(q/t)}{T_m(q/t)-\cos(\theta-\phi)} .$$

Hence, (20) becomes

$$F(t,\theta) = \frac{1}{2\pi^2 t} \int_0^{2\pi} d\phi \int_0^{\infty} \frac{\partial \hat{F}(q,\phi)}{\partial q} \frac{U_{m-1}(q/t)}{T_m(q/t)-\cos(\theta-\phi)} \, dq. \tag{21}$$

This contains Radon's original result in the special case $m = 1$, since $U_o = 1$ and $T_1(x) = x$. For the β-curves the result is identical except that the arguments of U_{m-1} and T_m are (t/q) instead of (q/t).

In conclusion, we briefly indicate an alternative approach to these special cases. If, in addition to making the radial scale change implied in (3), we make an angular scale change,

$$\theta = m\xi. \ \phi = m\eta \tag{22}$$

and if we define

$$F(s,\xi) = ms^{m-1}f(s^m,m\xi), \tag{23}$$

then $f(r,\theta)$ maps into the sector of the (s,ξ) plane given by $0 \leqslant \xi \leqslant \pi/m$. F can be smoothly and continuously extended to the rest of the (s,ξ) plane by the periodic condition

$$F(s,\xi + \pi/m) = F(s,\xi), \tag{24}$$

and, under (22) the equation of the α-curves becomes

$$s \cos(\xi-\eta) = q \tag{25}$$

the equation of a straight line in the (s,ξ) plane. If one solves Radon's problem ab initio (as for example in [2]) one obtains (5) with $\alpha = 1/m$, or if one starts with Radon's original solution ((21) with m=1) one arrives at (21)!

I am indebted to R.M Lewitt for drawing my attention to the work of Hansen and Verly on their extended circular harmonic transform.

BIBLIOGRAPHY

1. D. Behringer et al. (The Ocean Tomography Group) "A demonstration of ocean acoustic tomography", Nature 299 (1982) 121-125.

2. A.M. Cormack, "Representation of a function by its line integrals", J. App. Phys. 34 (1963) 2722-2727.

3. A.M. Cormack, "The Radon transform on a family of curves in the plane" I and II, Proc. Am. Math. Soc. 83 (1981) 325-330, 86 (1982) 293-297.

4. A.M. Cormack, "Computed Tomography: Some history and recent developments", Proc. Sym. in App. Math. 27 (1982) 35-42.

5. E.W. Hansen, "Image reconstruction from projections using circular harmonic expansion", Ph.D. Dissertation, Stanford University, Stanford, CA (1979).

6. E.W. Hansen, "Theory of circular harmonic image reconstruction", J. Opt. Soc. Am. 71 (1982) 304-308.

7. W. Magnus, F. Oberhettinger and R.P. Soni, Formulas and Theorems for the Special Functions of Mathematical Physics, Springer-Verlag, New York (1966).

8. A.A. Manuel, "Construction of the Fermi surface from positron-annihilation methods", Phys. Rev. Lett. 49 (1982) 1525-1528.

9. R.M. Perry, "Reconstruction of a function by circular harmonic analysis of its Radon transform", Digest of topical meeting on Image Processing for 2-D and 3-D Reconstruction, Opt. Soc. Am., Washington, DC (1975).

10. J.G. Verly, "Circular and extended circular harmonic transforms", J. Opt. Soc. Am. 71 (1981) 825-835.

PHYSICS DEPARTMENT
TUFTS UNIVERSITY
MEDFORD, MA 02155

SIAM-AMS PROCEEDINGS
Volume 14
1984

INVERSION OF THE X-RAY TRANSFORM

Kennan T. Smith

ABSTRACT. This article describes some mathematical aspects of computed tomography. The topics discussed are: uniqueness and nonuniqueness, stability, exact inversion formulas, approximate inversion formulas, Fourier transform formulas, and noise.

1. INTRODUCTION. The divergent beam x-ray transform of the function f from the source point a in the direction θ is defined by

$$(1.1) \qquad \mathscr{D} f(\theta) = \int^{\infty} f(a+t\theta)\,dt.$$

Physically, f is the x-ray attenuation function of the object x-rayed, a is the x-ray source, or tube focal point, and θ is an x-ray detector. $\mathscr{D} f(\theta)$ represents the attenuation in the x-ray beam along the ray with origin a and direction θ. The attenuation function f (commonly called the density function, to which it is similar) is assumed to be square integrable on R and to vanish outside a fixed bounded open set Ω. The problem of computed tomography is to reconstruct this function from a number of x-rays.

In a strict sense, the word tomography refers to the study of cross sections of f, i.e. to the 2-dimensional problem, and most of the activity in the field has centered on this case, but the 3-dimensional problem is beginning to attract more attention. To accomodate both the present discussion takes place in R .

A direction in R is a point of absolute value 1, i.e. a point on the unit sphere S . If $x \in R$, then $x^{\perp}$ is the subspace perpendicular to x.

Historically, computed tomography began with parallel beam x-rays in which the photons travel along lines with a fixed direction θ rather than

---

1980 Mathematics Subject Classification
Supported by the National Science Foundation under grant no. MCS-8101586.

along rays emanating from a fixed source a. The parallel beam x-ray transform is defined by

$$(1.2) \qquad \mathscr{P}_\theta f(x) = \int_{-\infty}^{\infty} f(x+t\theta)\,dt, \qquad x \in \theta^\perp.$$

Parallel beams are difficult to produce and are seldom used, but the formulas intertwine with those for the divergent beam and have historical interest as well.

Also of both historical and current interest is the Radon transform

$$(1.3) \qquad \mathscr{R}_\theta f(t) = \int f(x)\,dx$$

using the integrals of f over the planes perpendicular to a fixed direction θ. The very lively field of nuclear magnetic resonance involves the reconstruction of f from its Radon transform.

These mathematical transforms represent idealized approximations to the complex relationships between the object and the measured data. Discrepancies which result result from numerical approximation, positive diameter of the the x-ray sources and detectors, and incoherence of the x-ray beam, are not discussed in this report. The report deals mainly with the divergent beam transform, but includes occasional material on the others, mostly in the form of remarks.

2. UNIQUENESS AND NONUNIQUENESS. It is assumed that x-rays are taken from sources in some source set A outside the closed convex hull of Ω , and that for points a in A the attenuation is measured along rays with directions in a set S of directions. This section treats the extent to which the attenuation function is determined by the x-ray data.

(Normally A is a sphere surrounding Ω , but other configurations are of interest too. In dimension 2, S is usually the full circle S , but in dimension 3 there is coning to the region of interest.)

THEOREM 2.1 Let S = S . If A is any infinite set, then f is uniquely determined by the $\mathscr{D}$ f. Conversely, let A be any finite set. If K is any given compact set and g is any given function, then there exists a function f' with

$\mathscr{D}$ f' = $\mathscr{D}$ f for all a $\in$ A, and f' = g on K.

With coning there is no hope of uniqueness outside the region Ω consisting of those points x in Ω lying on at least one ray in one of the x-ray beams, i.e. x = a+t for some a $\in$ A and some $\theta \in$ S . Ω is called the measured region.

If Ω is the region of interest, the natural way to cone is to take S to be the set of directions θ such that the ray from a with direction θ meets Ω . Simple examples show that f is not uniquely determined on Ω (much less Ω') even when A is a full sphere surrounding Ω .

THEOREM 2.2 Let S = S $\subset$ S , and let $\overline{A}$, the closure of A, be a rectifiable arc. If for each $\theta \in$ S there is some a $\in \overline{A}$ such that the ray a+tθ misses $\overline{\Omega}$, then f is uniquely determined on Ω by the data $\mathscr{D}$ f(θ).

These results are proved in [6,8,17] along with similar results for the parallel beam and Radon transforms.

3. STABILITY. With a finite source set A = (a ,..., a) the operator

$$\mathscr{D} f = (\mathscr{D} f,\ldots,\mathscr{D} f)$$

is not one to one, but it nevertheless will have continuous pseudoinverses if the range is closed in a suitable topology on the range space.

To identify the suitable topology, note that for a fixed source a the Cauchy-Schwarz inequality gives

$$(3.1) \quad \int_S |\mathscr{D} f(\theta)| \, \delta (\theta) \, d\theta < \|f\| \quad \text{, where}$$

$$(3.2) \quad \delta (\theta) = \chi_\Omega (a+t)t \quad dt,$$

with χ_Ω the characteristic function of Ω . This suggests that the natural domain space for $\mathscr{D}$ is L (Ω), while the natural range space is the weighted L space L (δ), a suggestion confirmed by the fact that the adjoint, $\mathscr{D}$, becomes an isometry. Consequently, the natural range space for $\mathscr{D}$ is the product

$$(3.3) \quad L (\delta) \times \ldots \times L (\delta).$$

The operator $\mathscr{D}$ with domain L (Ω) is stable if its range is closed in the product (3.3).

The stability question was answered for very general Ω in dimension 2 and for convex Ω with smooth boundary in dimension 3 by D.V. Finch and D.C. Solmon [4]. To avoid the technical conditions in their theorem, it is assumed that Ω is convex with smooth boundary even in dimension 2.

THEOREM 3.4 If the boundary of Ω has positive Gaussian curvature at each point and if no line joining two sources is tangent to the boundary, then the range of $\mathscr{D}$ is closed. Both conditions are essentially necessary.

Stability of x-ray problems was investigated first for the Radon transform in dimension 2 [2,5,10,11], then in dimension n [3,12]. In [12] it was shown that the range is closed for any finite set of directions and for very general Ω . Instability appeared first in an example of the author showing the necessity of the non-tangency condition in the case of the 2 dimensional divergent beam and in the work of J. Boman [1] on the 3 dimensional parallel beam. Boman showed that for convex Ω with smooth boundary in R , the parallel beam range is closed for every finite set of directions if and only if the boundary has positive Gaussian curvature at each point. This work gave the first insights on instability and influenced strongly the subsequent work of Finch and Solmon.

Except in the "limited angle problem" (See M.E. Davison and F.A. Grunbaum [2]) stability has not played a decisive role in practice. The emphasis has been to find approximate inversion formulas for the case where the sources cover a full sphere around , then to implement the formulas with a finite number of sources. In the rest of the article A is a sphere with center 0 and radius R surrounding .

4. BACKGROUND FORMULAS. In this article the Fourier transform is given by

$$(4.1) \qquad \hat{f}(\xi) = \int e \qquad f(x)\,dx.$$

For use in potential theory M. Riesz [15] introduced the functions R_α defined by

$$(4.2) \qquad \hat{R}_\alpha(\xi) = |2\pi\xi|^{-\alpha}.$$

From the homogeneity and invariance under orthogonal transformations it follows that

$$(4.3) \qquad R_\alpha(x) = C(n,\alpha)\,|x|^{\alpha} \qquad \text{with } C(n,\alpha) = \qquad .$$

If the operator Λ is defined by

$$(4.4) \qquad (\Lambda f)^\wedge(\xi) = |2\pi\xi|\hat{f}(\xi)$$

then Λ^α inverts convolution by R_α. Λ is expressed directly by

$$(4.5) \qquad \Lambda f = \sum H * \partial f/\partial x \qquad \text{with } H(x) = (n-1)C(n,1)x\,/|x| \qquad .$$

the convolution being taken in the Cauchy principal value sense.

In addition to the x-ray and Radon transforms defined in section 1, two others will also play a role.

$$(4.6) \qquad \mathscr{L}f(y) = \int_{-\infty}^{\infty} f(x+t(y/|y|))dt, \qquad \tilde{\mathscr{L}}f(y) = |\langle y,x\rangle|\,|y|\ \ \mathscr{L}f(y).$$

The following relations are clear.

$$(4.7) \qquad \mathscr{L}\,f(\ ) = \mathscr{D}\,f(\ )+\mathscr{D}\,f(-\ ) = \mathscr{P}f(E\ x),$$

where E is the orthogonal projection on . Another important relation is

$$(4.8) \quad \int \mathscr{P}f(y)K(y)dy = (1/2R)\int_A \mathscr{L}f(\ )K(E\ a)da$$

where K is any function on . This is seen by making the change of variable y = E a, dy = |<a, >/R|da on the upper and lower hemispheres of A.

5. EXACT INVERSION FORMULAS. For a fixed source x the integral of f over the sphere S is given by

$$\int_S f(\ )d \quad = \quad \int_S \int^{\infty} f(x+t\ )dtd \quad = (1/C(n,1))R\ *f(x).$$

Therefore,

$$(5.1)\qquad f(x) = C(n,1)\quad \Lambda \int_S f(\ )d\ .$$

Replacing by - and adding the result to (5.1) gives

$$(5.2)\qquad f(x) = (1/2)C(n,1)\quad \Lambda \int_S {}^{-}f(\ )d\ .$$

Formula (4.7) then gives

$$(5.3)\qquad f(x) = (1/2)C(n,1)\quad \Lambda \int_S f(E\ x)d\ .$$

The corresponding Radon inversion formula is

$$(5.4)\qquad f(x) = (1/2)(2\ )\quad \Lambda \qquad \int_S \mathscr{R}\, f(\langle x,\ \rangle)d\ .$$

In formula (5.2), make the change of variable

$$= (a-x)/|a-x|, \text{ hence } d\ = (1/R)\,|\langle x-a,a\rangle|\,|x-a|\quad da.$$

When the homogeneity and symmetry

$$f(ty) = \quad f(y), \qquad f(a-x) = \quad f(x-a)$$

are used, formula (5.2) becomes

$$(5.5)\qquad f(x) = (C(n,1)/2R)\quad \Lambda \int_A f(x-a)da.$$

The exact inversion formulas for the x-ray and Radon transforms are formulas (5.3), (5.4), and (5.5). Formula (5.4) is due to Radon [13], and (5.3) probably has been known about equally long. Formula (5.5) and the simple proofs above appeared first in [8].

6. APPROXIMATE INVERSION FORMULAS. Because of the finite data, exact inversion cannot be achieved in practice, and indeed it is fruitful to adopt from the beginning the point of view that it is not a reconstruction of the attenuation function f that is sought, but rather an approximation

to it. Normally the approximation has the form e*f, where e is an approximate delta function, called the point spread function.

To obtain an approximate inversion formula for the parallel beam, replace f in (5.3) by e*f and use (e*f) = e* f to get

(6.1) $$e*f(x) = \int_S (\ f*k)(E\ x)d\ ,\ \text{with } k(y) = C(n,1)/2\,\Lambda\ \ e(y).$$

The corresponding formula for the divergent beam,

(6.2) $$e*f(x) = (1/2R) \int_S \int_A f(\ )k(E\ x-E\ a)dad\ ,$$

follows from (6.1) and (4.8) with K(y) = k(E x-y).

The parallel beam derivation also gives the approximate inversion formula for the Radon transform:

(6.3) $$e*f(x) = \int_S (\mathscr{R} f*k)(\langle x,\ \rangle)d\ ,\ \text{with } k(t) = (1/2)(2\ )\ \ \Lambda\ \ \mathscr{R}\, e(t).$$

In all of these cases the point spread function e can be chosen arbitrarily, and the kernel k derived from it. A few examples are given in [18]. This general form of the approximate inversion formulas appeared first in [8], except that (6.2) is a simple but important improvement of one of the formulas in [8]. In the case of dimension 2 and a particular point spread function e, (6.1) becomes the original parallel beam convolution formula of Ram and Lak [14], while (6.2) becomes the divergent beam formula of Lak [7]. (The latter is made clear in the derivation of H.J. Scudder [16].) The general case of (6.2) was noticed recently by F. Keinert.

Usually (6.2) is evaluated by integrating first with respect to . If x-a = r , the inner integral becomes

(6.4) $$\int_S f(\ )k(rE\ \xi)d\ ,$$

which involves excessive computation because of the presence of r = |x-a|

as an argument of k. One remedy is to replace $k(rE\,\xi)$ by r $k(E\,\xi)$, a heuristic justification for which is that k is an approximation to Λ δ, which is homogeneous of degree -n. With this approximation (6.2) becomes

$$(6.5) \qquad e*f(x) \sim (1/2R) \int_A |x-a| \int_S f(\;)k(E\,\xi)d\; da.$$

This and other approximations are discussed in H.J. Scudder [16] and R.M. Lewitt [9].

It is also possible to integrate first with respect to a. Let e be a radial point spread function, so that the kernel k is both radial and independent of . Identify R with the subspace x = 0 of R in the natural way. For each , let U be an orthogonal transformation of R carrying to (0,...,0,1). Then

$$(6.6) \qquad f*k(x') = (1/2R) \int_A f(\;)k(x'-U\;E\;a)da, \quad x' \quad R \quad ,$$

and (6.1) applies. This approach would appear to require a very large number of x-ray sources, with fewer detectors.

7. FOURIER TRANSFORM FORMULAS. It is immediately verified that the Fourier transform of f is the restriction of the Fourier transform of f to .

$$(7.1) \qquad (\;\;f)(\;\;) = f(\;\;) \qquad \text{for} \qquad .$$

For the divergent beam, the Fourier transform formula is

$$(7.2) \qquad f(\;\;) = (1/2R) \int_A e \qquad f(\;\;)da \;, \qquad ,$$

which follows from (4.8) with K(y) = e . For dimension 2, formula (7.2) was found by the author through calculation of the Fourier transforms of both sides of (5.5). The extension to dimension n and a simple proof were found by D.V. Finch. The formula appeared first in [18].

8. NOISE. The utility of mathematical formulas in practice depends heavily on their resistance to noise. In a recent coffee room conversation on

small parameter asymptotics in turbulence theory, P.G. Saffman remarked that the only true small parameter in turbulence theory is the signal to noise ratio. It is not the purpose of this section to discuss the nature of the noise in x-ray data, nor to discuss ways of handling it, but rather to present one example. The example consists of x-ray data and two reconstructions of a cross section of an exploded battery. The data were taken with a developmental version of an industrial scanner at the General Electric Research and Development Center by H.J. Scudder. The reconstruction on the left came from the raw data. The one on the right came from the same data subjected to a noise elimination procedure. This example is discussed in detail in [19], and an early version of the procedure is described in [17].

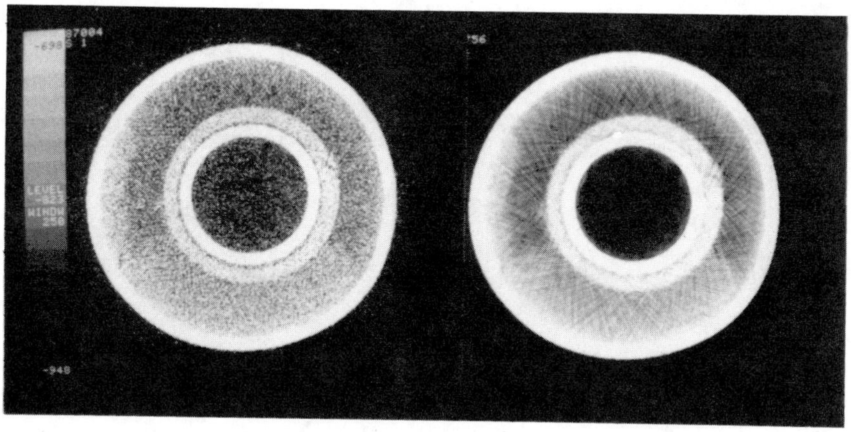

Cross sections of battery. The inner shell is steel, the adjacent shell is ceramic, and the outer shell is steel. The central region contains sodium mixed with less dense material, e.g. air bubbles. The region between the ceramic and outer steel shells contains an uneven mixture of sodium polysulfides. Before the explosion this region contained sulphur and the central region contained sodium.

The following graphs show typical x-ray data, the smoothed data, and a test of the noise elimination procedure on a known phantom to which noise from the x-ray data has been added.

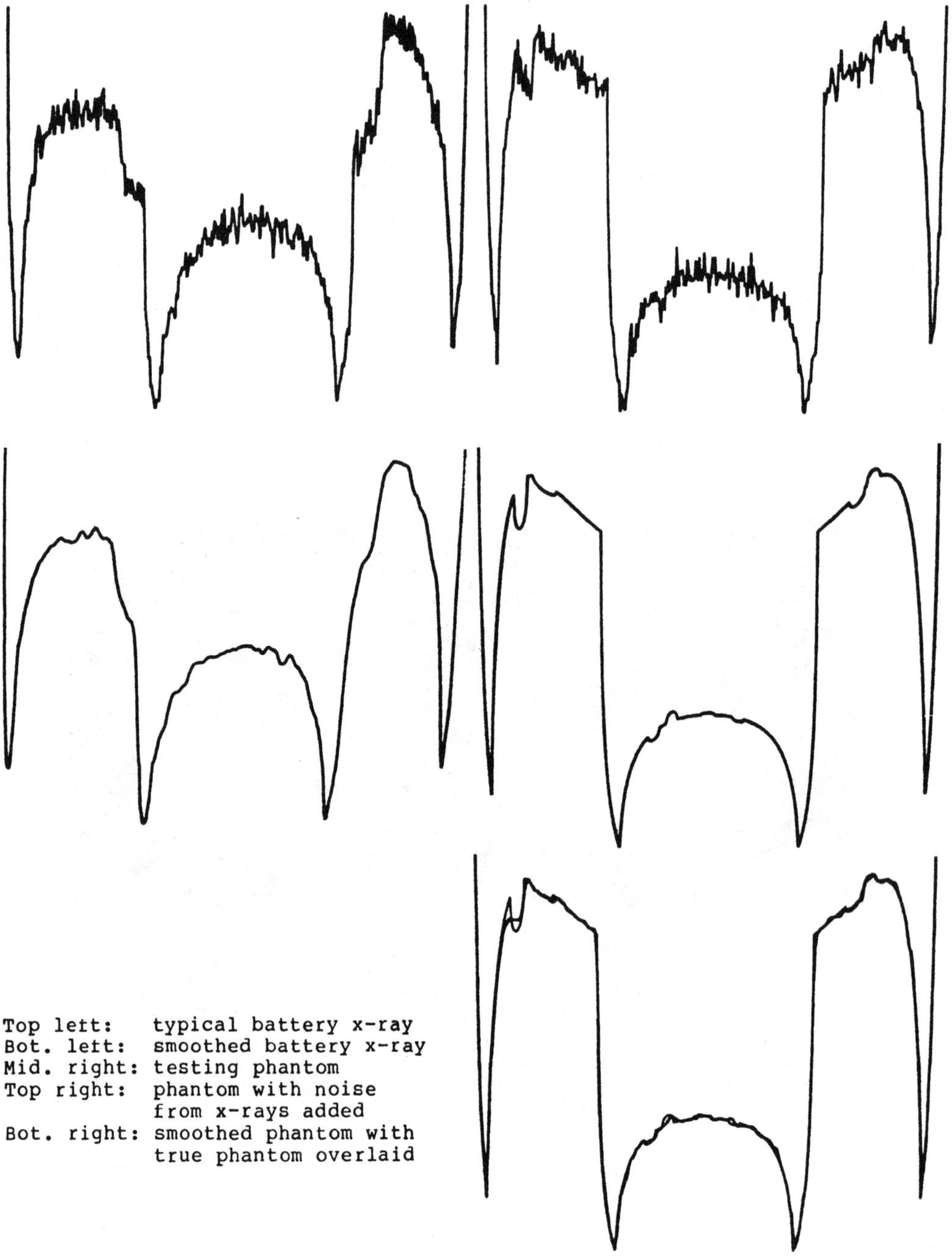

Top left: typical battery x-ray
Bot. left: smoothed battery x-ray
Mid. right: testing phantom
Top right: phantom with noise from x-rays added
Bot. right: smoothed phantom with true phantom overlaid

REFERENCES

1. J. Boman. On the closure of sums of plane waves and the range of the x-ray transform, I,II. Dept. of Mathematics, University of Stockholm 1981,1982. To appear: Ann. Inst. Fourier, Grenoble.

2. M.E. Davison and F.A. Grunbaum. Tomographic reconstruction with arbitrary directions. Comm. Pure. Appl. Math., 1981, 77-120.

3. K.J. Falconer. Consistency conditions for a finite set of projections of a function. Math. Proc. Cambridge Phil. Soc., 1979, 61-88.

4. D.V. Finch and D.C. Solmon. Sums of homogeneous functions and the range of the divergent beam x-ray transform. Num. Functional Anal. and Optim., 1983.

5. C. Hamaker and D.C. Solmon. The angles between the null spaces of x-rays. J. Math. Anal. Appl., 1978, 1-23.

6. C. Hamaker, K.T. Smith, D.C. Solmon, and S.L. Wagner. The divergent beam x-ray transform. Rocky Mt. J. Math., 1980, 253-283.

7. A.V. Lakshminarayanan. Reconstruction from divergent x-ray data. SUNY Tech. Report No. 92, Comp. Sci. Dept., Buffalo, NY, 1975.

8. J.V. Leahy, K.T. Smith, and D.C. Solmon. Uniqueness, nonuniqueness, and inversion in the x-ray and Radon problems. Internat. Symp. on Ill-Posed Problems, Univ. of Delaware, Newark, DE, 1979.

9. R.M. Lewitt. Reconstruction algorithms: transform methods. MIPG Tech. Rpt. No. MIPG74, Dept. of Radiology, U. of Penn., 1983.

10. B.F. Logan and L.A. Shepp. Optimal reconstruction of a function from its projections. Duke Math. J., 1975, 645-659.

11. R.B. Marr. An overview of image reconstruction. Internat. Symp. on Ill-Posed Problems, Univ. of Delaware, Newark, DE, 1979.

12. B.E. Petersen, K.T. Smith, and D.C. Solmon. Sums of plane waves and the range of the Radon transform. Math. Ann., 1979, 163-171.

13. J. Radon. Uber die Bestimmung von Functionen durch ihre Integralwerte langs gewissen Mannigfaltigkeiten. Ber. Verh. Sach. Akad. Wiss. Leipzig, Math.-Nat. kl., 1917, 262-277.

14. N. Ramachandran and A.V. Lakshminarayanan. Three-dimensional reconstruction from radiographs and electron micrographs: application of convolutions instead of Fourier transforms. PNAS USA, 1971, 2236-2240.

15. M. Riesz. Integrales de Riemann-Liouville et potentiels. Acta Szeged., 1938, 1-42.

16. H.J. Scudder. Introduction to computer aided tomography. Proc. IEEE, 1978, 628-637.

17. K.T. Smith, D.C. Solmon, and S.L. Wagner. Practical and mathematical aspects of reconstructing objects from radiographs. BAMS, 1977, 1227-1270.

18. K.T. Smith. Reconstruction formulas in computed tomography. Computed Tomography. Proc. Symp. Appl. Math. No. 27, L.A. Shepp, Ed., Amer. Math. Soc., Providence, RI, 1983.

19. K.T. Smith. Iterative noise elimination. Math. Res. Center Technical Report, University of Wis., 1983.

MATHEMATICS DEPARTMENT
OREGON STATE UNIVERSITY
CORVALLIS, OR 97331

Current Address:
Corporate Research and Development
General Electric Co. 37-217
Schenectady, NY 13345

SIAM-AMS PROCEEDINGS
Volume **14**
1984

COMPUTED TOMOGRAPHY FROM ULTRASOUND SCATTERED BY BIOLOGICAL TISSUES

James F. Greenleaf[1]

ABSTRACT. The inverse ultrasonic scattering problem was approached by discretizing the Rytov form of the Helmholtz equation and subsequently digitally backward propagating the measured scattered wave through the object space from a multiplicity of views. By summing the backward propagated phases, one obtains a first approximation to the refraction index distribution.

Advantages and disadvantages of the new method are discussed along with results in two dimensions.

1. INTRODUCTION. The development of computerized tomography in medical x-ray imaging has had a tremendous impact on the field of radiology (1). For the first time in the history of roentgenography, the radiologist can obtain quantitative measures of the spatial distribution of the linear attenuation coefficient for x-rays from a computed assisted tomography "CAT" scanner (2). Applications of computerized tomographic techniques are now being developed in many areas of medicine using a wide variety of energies such as x-radiation (3), electrons (4), gamma rays (5), alpha particles (6), nuclear magnetic resonance (7), positrons (8), and more recently ultrasound (9-18).

Ultrasound is a propagating disturbance of the material properties of the tissue through which it passes. It is currently the only nonelectromagnetic form of energy used for obtaining images in the body. Since the interactions of ultrasonic energy with tissue are very strong and complex, low levels of energy carry a large amount of information concerning the material properties of the tissue through which the energy has traversed. This strong interaction between tissue and ultrasound and the associated low level of energy, apparently allow ultrasound to be a safe and noninvasive imaging modality.

---

[1]Supported by Grants CA-24085 from the National Cancer Institute and ECS 7926008 from the National Science Foundation.

Current biomedical imaging methods are generally backscatter imaging methods. The B-scan is a naturally tomographic method which uses lenses to focus energy into the tissue, receives the backscattered energy and, through amplification and nonlinear mapping techniques, results in an image that gives a qualitative estimate of the backscatter strength within a plane through the object being scanned (19). Pathology is detected and diagnosed from the resulting images by characterizing the tissue using various morphologic parameters such as shadowing and by using alterations in visually appreciated texture in lesion areas and normal areas of the organ being imaged. Quantitative methods of evaluating backscatter coefficient or attenuation coefficients have been slow to develop because of the difficulty of the inverse scattering problem in the backscatter mode (20). Although some seismic techniques (21) and other linear integral techniques have been attempted, the field remains relatively undeveloped for measuring quantitative tissue parameters from the backscatter modes.

Ultrasonic computer-assisted tomography (UCAT) has resulted in quantitative measures of acoustic speed and some quantitative estimates of attenuation in those organs around which it is possible obtain the multiple views for acquiring data necessary for computed tomography (22,23). These observable organs are limited primarily to the head and breast.

Most of the transmission tomography methods developed for ultrasound utilize straight ray approximations to the propagation of waves in the tissue. Since waves propagate along curved rays and also undergo diffraction, the straight line approximation results in images which have aberrations that probably have an impact on the diagnostic capability of the modality (23).

Inverse scattering techniques based on the Born and Rytov approximations of the wave equation and the resulting inversions for the refraction index distributions from measurements obtained from multiple views around the object have been reported elsewhere (24) and experimental results have been recently obtained (25). These methods require an approximation to the wave equation to obtain a linear set of equations that can be solved analytically. It appeared to us that digitization of the wave equation itself, with no other approximations, would result in a better solution to the inverse problem than first linearizing the equation and then discretizing the resulting analytic solutions.

The purpose of this paper is to report a discretization technique for the wave equation which calculates scattering more precisely than the Rytov method and which can be used for calculating refraction indices in a manner analogous to that reported previously for the Rytov technique by Devaney, et al. (26).

2. METHODS. The Previous Method. The equation assumed to pertain during wave propagation in tissue for this study was the Helmholtz equation. This equation has inherent assumptions that the distribution of density is constant, i.e., has no partial derivatives, and that mode conversion, i.e., from compressional waves to surface or shear waves, is minimal. Under these assumptions, the wave equation can be written as the scalar Helmholtz equation if a single frequency of insonation is assumed, i.e.,

$$\nabla^2\Psi(\bar{r}) + k_o^2\eta(\bar{r})^2\ (\bar{r}) = 0, \quad (1)$$

where $k_o = \frac{w}{c_o}$, $w = 2\pi f$,

$\eta_{(\bar{r})} = \frac{c_o}{c(\bar{r})}$, $c(\bar{r})$ = Velocity,

and c_o = Reference Velocity.

Equation 1 can be transformed using the Rytov transformation in which the variable component of the pressure is placed in a complex phase term in the following way:

$$\Psi_o = \exp(\tilde{\gamma}), \text{ the Incident Wave} \quad (2)$$

and $\Psi = \exp(\gamma + \tilde{\gamma})$ the Total Wave, where Ψ, γ, and $\tilde{\gamma}$ are assumed to be functions of position $\bar{r}$.

Substituting (2) into the Helmholtz equation (Equation 1) gives

$$\nabla^2\gamma + 2\nabla\gamma\cdot\nabla\tilde{\gamma} + \nabla\gamma\cdot\nabla\gamma + k^2(\eta^2-1) = 0. \quad (3)$$

If we assume that $\nabla\gamma \ll \nabla\tilde{\gamma}$, (4)

the first Rytov approximation becomes

$$\nabla^2\gamma + 2\nabla\gamma\cdot\nabla\tilde{\gamma} + k^2(\eta^2-1) = 0. \quad (5)$$

Substituting 4 into 5 and defining $\phi_s = \Psi_o\gamma$, gives the following constant coefficient inhomogeneous Helmholtz equation,

$$\nabla^2\phi_s + k^2\phi_s = -k^2(\eta^2-1)\Psi_o. \tag{6}$$

This equation is analytically solvable using Green's functions, given appropriate boundary conditions (24).

The New Method. Rather than obtain an analytic solution for Equation 6, we begin with Equation 3, assume that plane wave insonation was used and then discretize Equation 3 on a square N x N grid, for example, resulting in Equation 7 for the elements of the grid.

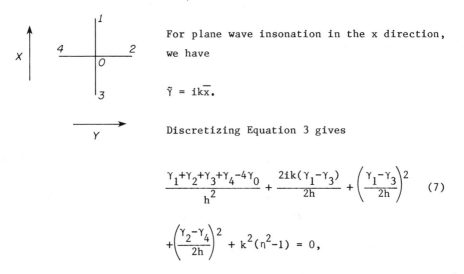

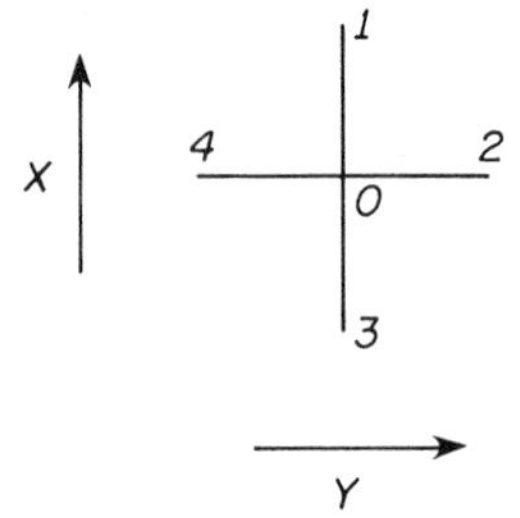

For plane wave insonation in the x direction, we have

$$\tilde{\gamma} = ik\bar{x}.$$

Discretizing Equation 3 gives

$$\frac{\gamma_1+\gamma_2+\gamma_3+\gamma_4-4\gamma_0}{h^2} + \frac{2ik(\gamma_1-\gamma_3)}{2h} + \left(\frac{\gamma_1-\gamma_3}{2h}\right)^2 \tag{7}$$

$$+\left(\frac{\gamma_2-\gamma_4}{2h}\right)^2 + k^2(\eta^2-1) = 0,$$

where γ_n represents the complex phase at the relative grid points shown in the diagram, where h is the grid step size.

Given that we have starting values for $\gamma(x,y)$ $x=x_o+nh$ $n=o,1$ and $y=y_o+nh$ $n=o,\ldots,N-1$, i.e., for γ_o, γ_2, γ_3, γ_4, then we can solve for $\gamma_1= \gamma(x_o+2h,y)$, for $y=y_o+nh$ $n=o,\ldots,N-1$, and so on incrementing in the x direction until the entire grid is finished. For forward propagation, the complex wave number k of the incident wave must be known.

For backward propagation, we assume that the phase of the scattered wave $(\gamma+\tilde{\gamma})$ has been measured at $y=y_o+nh$ $n=1,\ldots,N-1$ and for the line $x=x\ell$ (Figure 1). We then assume that the wave is the same at the line $x=x\ell-h$. Given these two lines we then can backward propagate by solving Equation 7 for $\gamma_3= \gamma(x-2h,y)$, $y=y_o+nh$ $n=1,\ldots,N-1$ as before, one line at a time.

Errors in this assumption are small if the step size h is small and if the wave is propagating essentially normal to the y axis.

3. RESULTS. In order to determine the accuracy of the forward scattering predicted by the discretized wave Equation 7, we placed a condom filled with an alcohol/water mixture in a water bath, propagated a plane wave through the condom and measured the scattered energy. The speed of sound in water relative to that in the "phantom" was 1.026 and the wave number k was 2.85/mm. Figure 1 illustrates the geometry of this experiment. The plane wave propagated orthogonal to the axis of the object, and the scattered wave was measured with a small hydrophone which was scanned along a locus a fixed distance from and parallel to the plane wave transmitting transducer.

The results are shown in Figure 2 in which the amplitude of the scattered wave is compared to the scattered wave calculated by the Rytov method (Equation 5) and by the discretized wave equation (Equation 7). One can see that the discretized wave equation is much more accurate than the Rytov approximation.

4. RECONSTRUCTION (INVERSE SCATTERING). Devaney, et al. (11) have developed a backward propagation technique for reconstruction of the refractive index. The Devaney method is to 1) measure the scattered wave from plane wave insonation at many angles of view ϕ_k k=1,...,K. 2) At each angle of view the measured scattered wave is backward propagated across the region of interest using equations developed from the Rytov approximation. 3) By summing all of the backward propagated complex phases for all views one obtains an approximation of the refractive index distribution in the region.

We have developed a reconstruction method analogous to that of Devaney. However, instead of using the Rytov approximation, we use the discretized equation (Equation 7). We reconstructed the refractive index by backward propagating from a multiplicity of views and summing the phases from all those views, in analogy to the backward propagation methods of Devaney.

Figure 3 illustrates a comparison of the reconstructed distributions of refractive indices calculated using the Rytov approximation and the discretized wave equation. The algorithm was to 1) calculate to the wave scattered from a circular symmetric and centered disc having the same characteristics as the alcohol and water filled phantom, 2) backward propagate the calculated scattered wave through the object region assuming that the refraction index within the object region was constant, 3) repeat the

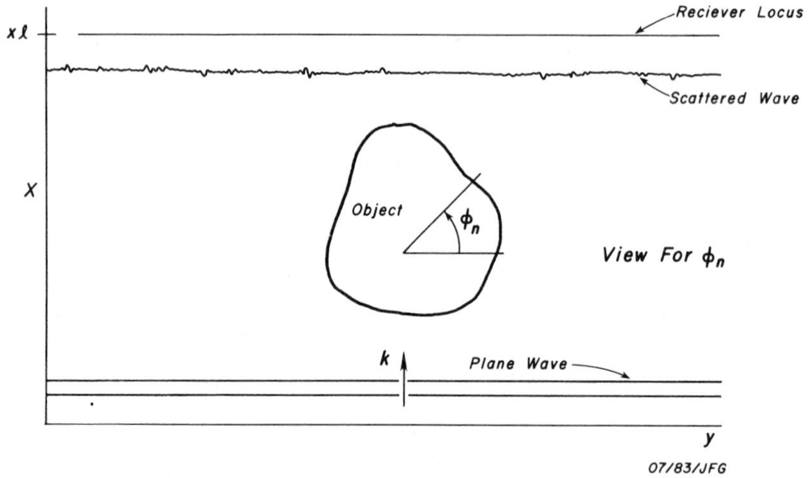

Figure 1 Geometry of diffraction tomography problem. For complete reconstruction the object must be rotated through many angles of view and data collected at $x = x\ell$ for each view.

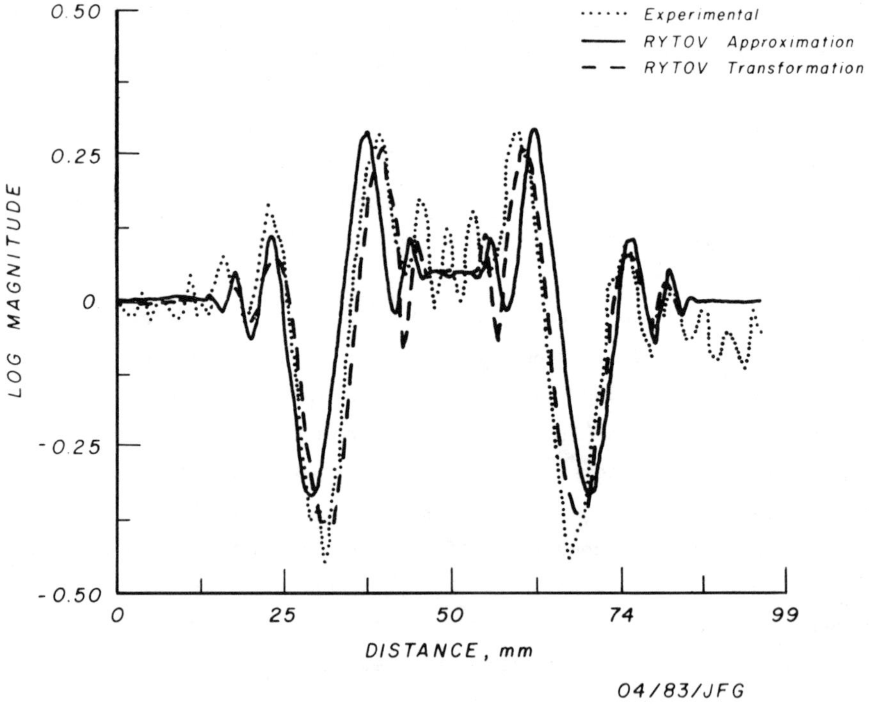

Figure 2 Comparison of scattered wave measured at 32 mm from the center of a cylindrical scatterer of radius 18 mm. Refractive index n = 1.026, wave number k = 2.85/mm. Note the experimental profile most closely resembles that calculated from the Rytov transformation (derived from Equation 7) rather than from the Rytov approximation derived by ignoring nonlinear terms in Equation 7.

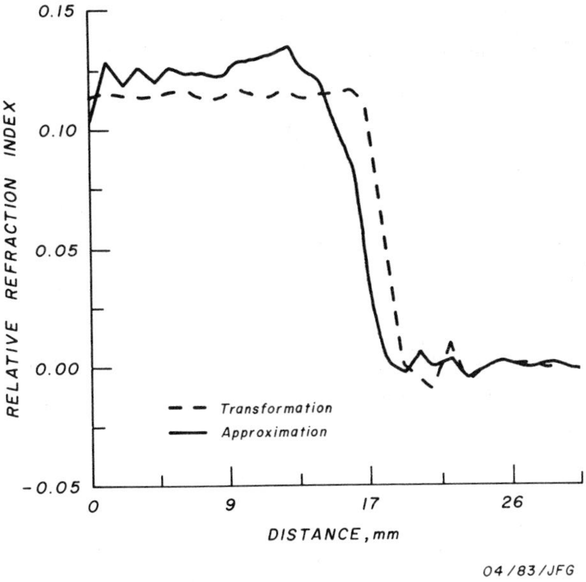

Figure 3 Comparison of reconstructions from simulated scattered profiles shown in Figure 2 using the backward propagation method of Devaney. The Rytov approximation (---) is compared with the new discretized wave equation method (——). The radius of the cylinder was 18 mm.

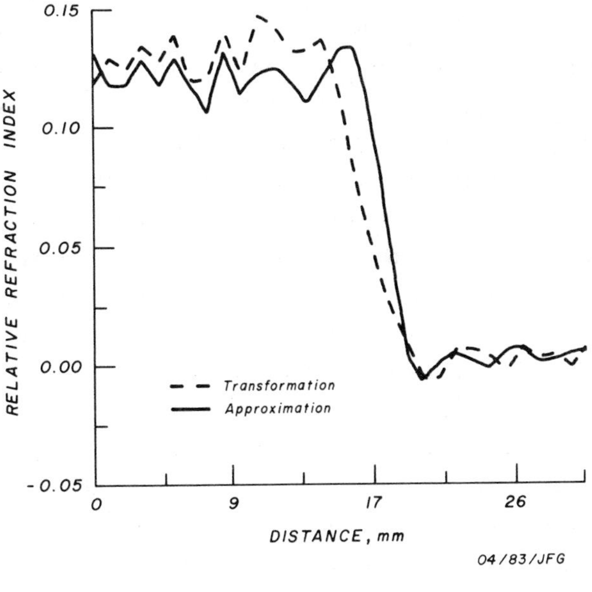

Figure 4 Comparisons of reconstructions from measured profiles using the Rytov --- and discretized wave equation (——) methods (radius = 18 mm, $n = 1.026 + i0$, $x\ell = 32$ mm, $k = 2.85$/mm).

backward propagation from a series of 47 equispaced angles of view, and 4) sum the phases of the backward propagated waves resulting in the reconstruction of the refraction index shown in Figure 3.

The same algorithm was applied to experimental data measured from the wave scattering off of the alcohol and water filled phantom. The resulting profiles through the reconstructed cylinder are shown in Figure 4. One can see that the reconstructed size of the object is more accurate for the discretized wave equation reconstruction than for the Rytov approximation. This seems apparent from Figure 2 since the Rytov approximation does not accurately predict the position of the edges of the scattered wave, thus backward propagation using the same algorithm does not predict accurately the size of the object.

5. DISCUSSION. The disadvantage of these reconstruction techniques is that the complex phase must be calculated from measured pressure distributions. This requires one to obtain the complex logarithm of the measured data so that the phase must be known absolutely rather than modulo 2π. This requires phase unwrapping which, in complex cases, can be extremely difficult. We currently feel that the major obstacle to using these techniques is the phase unwrapping problem. A drawback to solving the discretized wave equation technique is that it is relatively slow. However, implementation on array processors may make it as virtually as fast as the backward wave propagation of Devaney. Rolf Mueller, et al. (27) have demonstrated that a Fourier transform reconstruction technique for the Rytov method can be implemented which is much faster than the backward propagation technique and which is as accurate.

This method can be applied to three-dimensional scattering by adding two elements to the stencil of Equation 7. Then one would backpropagate a plane rather than a line. Thus, the mathematical complexity only increases by about 2/5 while the requirements for storage increase by a factor of N. The discretized wave equation propagation reconstruction method could also be used in an iterative mode in which a current estimate of the refractive index is estimated from a previous reconstruction. In this case, the measured wave can be backward propagated through the current estimate of the refractive index and a higher order of approximation can be obtained. This cannot be done with the analytic inversion techniques such as the Fourier transform Rytov inversion method or the backward propagation method of inversion.

One also might consider backward propagating the pressure using the Helmholtz equation and Fourier transform methods described elsewhere (21); however, the resulting distribution of pressure must then be phase unwrapped in order to solve for the complex refractive index. This results in a two-dimensional phase unwrapping problem which may be more difficult than the one-dimensional phase unwrap problem.

Whether these methods can be applied in vivo to the problem of detecting breast cancer is yet to be determined.

Acknowledgments. The author appreciates the help of Elaine Quarve for typing the manuscript and Steve Richardson and James Hanson for graphic assistance.

BIBLIOGRAPHY.

1. See for instance the J Comput Assist Tomogr, Raven Press.

2. R. A. Brooks and G. DiChiro, Principles of computer assisted tomography (CAT) in radiographic and radioisotopic imaging, Phy Med Biol 21 (1976), 689-732.

3. A. M. Cormak, Representation of a function by its line integrals, with some radiological applications, J Appl Phys 34 (1963), 2722-2731.

4. D. J. DeRosier and A. Klug, Reconstruction of three dimensional structures from electron micrographs, Nature (London) 217 (1968), 130-134.

5. D. E. Kuhl, J. Hale, and W. L. Eaton, Transmission scanning: A useful adjunct to conventional emission scanning for accurately keying isotope deposition to radiological anatomy, Radiology 87 (1966), 278-284.

6. K. M. Crowe, T. F. Budinger, J. L. Cahoon, V. P. Elischer, R. H. Huesman, and L. L. Kanstein, Axial scanning with 900 meV alpha particles, IEEE Trans Nucl Sci NS-22(3) (1975), 1752-1754.

7. P. Lauterbur, Magnetic resonance zeugmatography, Pure Appl Chem 40 (1974), 149-157.

8. M. M. Ter-Pogossian, M. E. Phelps, E. J. Hoffman, and N. A. Mullani,[1] A positron-emission transaxial tomography for nuclear imaging (PETT)[1], Radiology 114 (1975) 89-98.

9. J. F. Greenleaf, S. A. Johnson, S. L. Lee, G. T. Herman, and E. H. Wood, Algebraic reconstruction of spatial distributions of acoustic absorption with tissues from their two-dimensional acoustic projections. In: P. S. Green: Acoustical Holography, New York, Plenum Press, 5 (1974), pp 591-603.

10. J. F. Greenleaf, S. A. Johnson, W. F. Samayoa, and F. A. Duck, Algebraic reconstruction of spatial distributions of acoustic velocities in tissue from their time-of-flight profiles. In: N. Booth: Acoustical Holography, New York, Plenum, 6 (1975), pp 71-90.

11. J. F. Greenleaf, and S. A. Johnson, Algebraic reconstruction of spatial distributions of refractive index and attenuation in tissues from time-of-flight and amplitude profiles. Proc Seminar Ultrason Tissue Characterizations, Natl Bur Stand Special Publ 453 (1975), 109-119.

12. S. A. Johnson, J. F. Greenleaf, C. R. Hansen, W. F. Samayoa, M. Tanaka, A. Lent, D. A. Christensen, and R. L. Woolley, Reconstructing three-dimensional fluid velocity vector fields from acoustic transmission measurements. In: L. Kessler: Acoustical Holography, New York, Plenum, 7 (1977), 307-326.

13. J. F. Greenleaf, S. A. Johnson, W. F. Samayoa, and C. R. Hansen, Refractive index by reconstruction: Use to improve compound B-scan resolution. In: L. Kessler: Acoustical Holography, New York, Plenum, 7 (1977), pp 263-273.

14. G. H. Glover, Characterization of in vivo breast tissue by ultrasonic time-of-flight computed tomography, Natl Bur Stand Int Symp Ultrason Tissue Characterization, National Science Foundation, Ultrasonic Tissue Characterization II, June 13-15, 1977 (1979), pp 221-225.

15. P. L. Carson, L. Shabason, D. E. Dick, and W. Clayman, Tissue equivalent test objects for comparison of ultrasound transmission tomography by reconstruction with pulse echo ultrasound imaging, Natl Bur Stand Int Symp Ultrason Tissue Characterization, National Science Foundation, Ultrasonic Tissue Characterization II, June 13-15, 1977 (1979), pp 337-340.

16. P. J. Carson, T. V. Oughton, W. R. Hendee, and A. S. Ahuja, Imaging soft tissue through bone with ultrasound transmission tomography by reconstruction, Med Phys 4 (1977), 302-309.

17. G. H. Glover and L. C. Sharp, Reconstruction of ultrasound propagation speed distributons in soft tissue: time-of-flight tomography, IEEE Trans Sonics Ultrason SU-24(4) (1977), 229-234.

18. R. Dunlap, Ultrasonic CT imaging, Ph.D. Thesis, Christchurch, New Zealand (1978).

19. M. Fatemi and A. C. Kak, Ultrasonic B-scan imaging: Theory of image formation and a technique for restoration, Ultrasonic Imaging 2(1) (1980), 1-47.

20. F. A. Duck and C. R. Hill, Acoustic attenuation reconstruction from backscattered ultrasound. In: J. Raviv, J. F. Greenleaf, and G. T. Herman: Computer Aided Tomography and Ultrasonics in Medicine, Amsterdam, North-Holland (1979), pp 137-150.

21. Norton, S. J. and M. Linzer, Ultrasonic reflectivity imaging in three dimensions: Reconstruction with spherical transducer arrays, Ultrasonic Imaging 1 (1979), pp 210-231.

22. P. L. Carson, C. R. Meyer, A. L. Scherzinger, and T. V. Oughton, Breast imaging in coronal planes with simultaneous pulse echo and transmission ultrasound, Science 214 (1981), 1141-1143.

23. J. F. Greenleaf, Computerized transmission tomography. In: P. D. Edmonds: Methods of Experimental Physics-Ultrasound, New York, Academic Press, 19 (1981), pp 563-589.

24. R. K. Mueller, Diffraction tomography I: The wave equation, Ultrason Imaging 2 (1980), 213-222.

25. M. Kaveh, M. Soumekh, Z. Q. Lu, R. K. Mueller, and J. F. Greenleaf, Further results on diffraction tomography using Rytov's approximation. In: E. A. Ash and C. R. Hill: Acoustical Imaging, New York, Plenum Press, 12 (1982), 599-608.

26. A. J. Devaney, A filtered backpropagation algorithm for diffraction tomography, Ultrasonic Imaging 4 (1982), 336-350.

27. R. K. Mueller, Department of Engineering, University of Minnesota, Personal Communication.

BIODYNAMICS RESEARCH UNIT
DEPARTMENT OF PHYSIOLOGY AND BIOPHYSICS
MAYO FOUNDATION
ROCHESTER, MN 55905

SIAM-AMS PROCEEDINGS
Volume **14**
1984

SOME MATHEMATICAL PROBLEMS SUGGESTED BY LIMITED ANGLE TOMOGRAPHY

F. Alberto Grünbaum

Table of Contents

1. FOREWORD

The intention of this lecture is to formulate for the non-expert a version of the "limited angle" problem as it appears in tomography. We plan to show a) what sort of mathematics one is led into, in trying to analyze the problem, b) what sort of new mathematical questions are suggested along the way.

There are quite a number of sources the reader can consult to get a more systematic view of the many fascinating mathematical and practical problems posed by tomography [1, 2, and their references].

Before getting down to business it might be appropriate to state here the main issue for this lecture: what is the trade-off between "range of views" or "radiation dosage" and "picture quality"?

Put in more dramatic terms: does a saving of 10% in data collection time reduce the picture quality by 5%, 40%, or 90%?

As anyone can imagine even the task of giving these terms a precise meaning in any imaging problem is not beyond debate. It turns out that even if one settles for a precise (if not universally acceptable) notion of

---

The research reported here was partially supported by National Science Foundation Grant MCS81-07086.

"picture quality" like "the number of essentially non zero singular values of a certain linear operator," one is usually left with an impossible computational problem.

Now a miracle comes into play. For the case of limited angle tomography the computation of singular vectors and values can be carried out when the range of views is a connected arc $|\theta| \le \alpha$. This miracle is the take-off point for this lecture. One would like to extend this to other situations: I have failed at this. Given this, one would like to understand what is behind the successful case. The last sections in the lecture are a beginning in that direction.

2. MATHEMATICAL SETUP

Let D be the unit disk on R^2. The same considerations apply in R^3 if line integrals are replaced by plane integrals, but we restrict ourselves to the two dimensional case in this section.

Let f, the unknown, be supported in D and belong to $L^2(D, w_2)$ with $w_2(r) = (1 - r^2)^{1-\lambda}$ $\lambda > 0$ arbitrary.

Define the "projections"

$$(P_\theta f)(t) = \int_{x \cos\theta + y \sin\theta = t} f, \quad t \in R.$$

Problem. Find f from the data $P_\theta f$, $|\theta| \le \alpha$. We say that we have a "limited angle" problem when $\alpha < \pi/2$.

Let I denote the interval $[-1, 1]$ and $w_1(s) = (1 - s^2)^{1/2-\lambda}$. One can see that $(Pf)_\theta \in L^2(I, w_1)$ and thus one can formulate the problem as

$$Pf = \text{data}$$

with $P : L^2(D, w_2) \to \oplus L^2(I, w_1)$. The symbol $\oplus$ denotes a continuous direct sum.

The standard way to handle such a problem is to look at the maps P^*P or PP^*, and find their eigenvectors and eigenvalues. The degree of ill-conditioning of the reconstruction problem at hand is given by this spectral information since the "best" solution is given by

$$f = \Sigma 1/_{\lambda i} \, (\text{data}, \phi_i) \, \varphi_i$$

where φ_i, ψ_i are the eigenvectors of P^*P, PP^* respectively, and λ_i the square roots of their (non negative) eigenvalues.

The point of this formula is that it shows clearly where the troubles will be coming from: small λ_i are sources of trouble unless f has no component along the corresponding φ_i, otherwise errors in the data if they happen to have components along ψ_i propagate catastrophically. In summary one needs not only the λ_i's but also the φ_i's and ψ_i's.

For most linear problems this is the end of the line. You cannot find the spectral information $\{\varphi_i, \psi_i, \lambda_i\}$ even if you are satisfied with accurate numerical approximations.

But in our case things turn out fine.

We consider only the problem of finding $\{\psi_i, \lambda_i\}$, the search for φ_i is analogous.

The map PP^* takes $\oplus L^2(I, w_1)$ into itself, and we have a decomposition

$$\oplus L^2(I, w_1) = \sum_{n=0}^{\infty} V_n$$

into a discrete sequence of infinite dimensional subspaces V_n given by

$$V_n = \{\varphi(\theta, s) \mid \varphi = g(\theta) \frac{\mathbb{C}_n^\lambda(s)}{w_1(s)}\} .$$

Here $C_n^\lambda(s)$ denote the Gegenbauer polynomials and $g(\theta)$ is arbitrary (in an appropriate Hilbert space).

The point of this decomposition is that PP^* leaves each V_n invariant, and its action on it, is given by

$$((PP^*)_n g)(\theta) = \int_{-\alpha}^{\alpha} \frac{\mathbb{C}_n^\lambda(\cos(\theta - \theta'))}{C_n^\lambda(1)} g(\theta')d\theta', \ |\theta| < \alpha$$

Notice that on each V_n, PP^* acts as a finite convolution integral operator. Once the eigenfunctions of $(PP^*)_n$, denoted by $H_{n,k}(\theta)$, $k = 0, 1, 2, \ldots,$ are found we obtain the eigenfunctions of PP^* in the form of products

$$H_{n,k}(\theta) \frac{\mathbb{C}_n(s)}{w_1(s)} .$$

The miracle comes in at this point. The integral operator given above allows for a differential operator which commutes with it, namely

$$\frac{d}{d\theta}((\cos\theta - \cos\alpha)\frac{d}{d\theta}) + (n^2 - 1)\cos\theta .$$

Since this latter operator has simple spectrum in $L^2[-\alpha, \alpha]$ each one of its eigenfunctions is an eigenfunction of the integral operator.

From the numerical point of view the computation of (a large number of) the eigenfunctions of a differential operator is a much more manageable task than the same job for an integral operator: one could use just the last steps of the Q - R algorithm, and you don't have to handle a very large full matrix.

The reader interested in a more detailed version of this development should consult references 3 through 10.

3. A MODEL PROBLEM

The limited angle problem discussed in the previous section has a unique solution given by analytic continuation. This follows from the assumption that f has compact support and the fact that the data at hand is equivalent to the knowledge of $\hat{f}(\Lambda_1, \Lambda_2)$ on the "wedge" $|\lambda_2/\lambda_1| \leq \tan\alpha$. This uniqueness accounts for the fact that all λ_i's in the last section are positive. The well known fact that analytic continuation is a very ill-conditioned problem shows up in the fact that except for a few values of i, the λ_i's are "essentially" zero.

We illustrate this with a well known problem; the analogy with our "limited angle" problem is apparent.

<u>Problem</u>. If f has support in $[-T, T]$ and lies in $L^2[-T, T]$, find f from the data $\hat{f}(\lambda)$, $|\lambda| \leq \Omega$.

For illustration purposes we consider a popular iteration algorithm to handle this problem [11, 12, 13, 14]. The algorithm proceeds by a sequence of transformations between physical and Fourier space; at each stage one corrects the data according to the a priori information: $\hat{f}(\lambda)$ is known for $|\lambda| \leq \Omega$ and f has support in $[-T, T]$. Although one can "prove" the convergence of this algorithm for each f, in practice one finds that the iterates $f^{(n)}$ converge to the projection of f onto a subspace of $L^2[-T, T]$ which depends on Ω and is described as follows:

The integral operator

$$(Kg)(x) = \int_{-T}^{T} \frac{\sin \Omega(x - y)}{x - y} g(y)dy \quad |x| \leq T$$

has discrete spectrum

$$1 > \lambda_1 > \ldots > \lambda_s > \lambda_{s+1} > \ldots > 0$$

with (typically) $\lambda_s \cong 0.3$, $\lambda_{s+1} \cong 10^{-8}$ and $s \cong [4\Omega T]$. The corresponding eigenfunctions are given by

$$\varphi_1, \varphi_2, \ldots, \varphi_s, \varphi_{s+1}, \ldots .$$

The subspace in question is the linear span of $\varphi_1, \ldots, \varphi_s$.

Indeed if we put

$$f = \sum_{i=1}^{\infty} \mu_i \varphi_i$$

one proves that the n-th iterate of algorithm gives

$$f^{(n)} = \sum_{i=1}^{\infty} \mu_i (1 - (1 - \lambda_i)^n) \varphi_i .$$

This expression shows that in practice

$$f^{(n)} \to \sum_{i=1}^{s} \mu_i \varphi_i .$$

For a given Ω (assuming T fixed) one would like to determine $\{\lambda_i, \varphi_i\}$ to decide what functions f will be possible to reconstruct from $\hat{f}(\lambda)$, $|\lambda| \leq \Omega$.

Here another miracle pops up. The integral operator K given above commutes with the differential operator

$$\frac{d}{dx}((T^2 - x^2)\frac{d}{dx}) - x^2\Omega^2$$

and we are in the situation described earlier [6, 7, 8].

The reader should be warned that miracles do not abound. If we replace the range $|\lambda| \leq \Omega$ by any set $\tilde{\Omega}$ — with some abuse of language — we get an integral operator whose kernel becomes

$$\int_{\tilde{\Omega}} e^{i\lambda x} e^{-i\lambda y} d\lambda$$

instead of the kernel $\sin \Omega(x - y)/(x - y)$.

It is a fact that unless $\tilde{\Omega}$ is an interval centered at the origin, the integral operator in question does not allow for any differential operator which commutes with it.

It would be nice to relax this commutativity requirement while preserving a large set of common eigenvectors for both operators but we have not succeeded in this.

Rather we turn to an investigation of the "reasons" for the miracle, in the cases when it works.

4. COMMUTING INTEGRAL AND DIFFERENTIAL OPERATORS

We record here a number of integral operators built up in exactly the manner discussed earlier, except that now we replace the real line by a different space. In particular all these examples substitute $[-T, T]$ by a "ball" of points close to the origin, and $[-\Omega, \Omega]$ is replaced by some kind of "low pass filter" whereby only low frequencies are preserved. For details, see [2] and its references.

a) Replacing R by R^n, build the kernel of the integral operator out of the eigenfunctions of the Laplacian in R^n. For simplicity restrict attention the radial functions in which case we are looking at the eigenfunctions

$$\varphi(\lambda, x) = \sqrt{\lambda x}\, J_{(n-2)/2}(\lambda x)$$

of the Laplacian

$$-\frac{d^2}{d^2x^2} + \left(\frac{n-1}{2}\right)\left(\frac{n-1}{2} - 1\right)\frac{1}{x^2} .$$

In this case one can find a differential operator which commutes with the integral operator built out of these Bessel functions.

b) Replacing R by S^n, the n-dimensional sphere, we build the integral operator out of the eigenfunctions of the Laplacian in

S^n. For simplicity, again, restrict attention to functions depending only on colatitude θ, and you are then looking at the eigenfunctions of

$$-\frac{d^2}{d\theta^2} + \frac{((n-1)/2)((n-1)/2-1)}{\sin^2\theta} .$$

In this case everything goes fine once again. Indeed the case of $n = 1$ is exactly the example discussed in section 2.

Notice that the eigenfunctions used to build up the integral operator are Gegenbauer polynomials; for instance if $n = 2$ we get Legendre polynomials

$$\varphi(k, \theta) = P_k(\cos\theta).$$

This two dimensional example comes up in NMR tomography and was the starting point of all these developments [15].

c) Replacing R by H^n, the n-dimensional hyperbolic space sitting in R^{n+1}, we end up looking at the eigenfunctions of the operator

$$-\frac{d^2}{dr^2} + \left(\frac{((n-1)/2)((n-1/2)-1)}{\sinh^2 r} + \frac{(n-1)^2}{2}\right) .$$

These eigenfunctions can be written down in terms of modified Legendre functions.

For $n = 3$ one gets a drastic simplification, namely

$$\varphi(\lambda, r) = \frac{\sin\lambda r}{\lambda\sinh r} .$$

In this case $n = 3$ we succeeded in proving that the "miracle" works. Later we managed to prove that this is the <u>only</u> value of n with this property [16].

d) Abandon the geometrical setup and consider the eigenfunctions of the Hermite operator

$$-\frac{d^2}{dx^2} + x^2 .$$

Denote the eigenfunctions by $h_n(x)$. Then one can prove that for each X, N the matrix

$$G_{ij} \equiv \int_{-\infty}^{X} h_i(\xi)h_j(\xi)d\xi \qquad 0 \leq i, j \leq N$$

commutes with a tridiagonal matrix—which can be written down very explicitly—with simple spectrum. We contend that this is the natural analog of the "miracle." This property holds true if we start with any of the "classical orthogonal polynomials": Jacobi, Laguerre, Bessel, Hahn, Krawtchouk, Askey-Wilson, ... On the other hand the property fails on any example that we have considered thus far and which does not come from this list [17].

A look at the examples mentioned above suggests that what may be lurking behind the "miracle" is the fact that the eigenfunctions $\varphi(\lambda, r)$ should satisfy some nice differential equation in the spectral parameter λ. Indeed in all the examples above where the "miracle" works, this property holds with a second order operator in λ.

This is the motivation for the next and final section.

5. RELATED PROBLEMS

Section 4 should provide enough motivation for the following pair of problems.

a) Find "potentials" $V(x)$ such that for some family of eigenfunctions $\varphi(x, \lambda)$ of the Schroedinger equation

$$\left(-\frac{d^2}{dx^2} + V(x)\right)\varphi(x, \lambda) = \lambda^2\varphi(x, \lambda)$$

we have an equation of the form

$$\sum_{i=0}^{M} a_i(\lambda)\frac{\partial^i\varphi}{\partial\lambda^i} = \theta(x)\varphi(x, \lambda) .$$

b) Find potentials $V(x)$ such that for some family of eigenfunctions $\varphi_n(x)$ we get an equation

$$\sum_{i=-M}^{M} a_i(n)\varphi_{n+i}(x) = \theta(x)\varphi_n(x) .$$

The initial work on question a) was done in collaboration with H. Duistermaat who managed to find a neat algebraic necessary condition on V. Subsequent work was done by the author, but it has not reached yet a definite state [18]. However we consider the examples below of enough interest to be presented at this time.

Example 1. Take the potential

$$V(x, t) = \frac{6x(x^3 - 2t)}{(x^3 + t)^2}$$

with eigenfunctions given by

$$\varphi(x, \lambda) = e^{i\lambda x} \frac{\theta_2(x - 1/(i\lambda), t + 1/(i\lambda)^3)}{\theta_2(x, \lambda)}$$

with $\theta_2(x, t) = x^3 + t$.

Then we have the differential equation

$$(-\frac{\partial^2}{\partial\lambda^2} + \frac{6}{\lambda^2})^2\varphi - 4ti \frac{\partial}{\partial\lambda}\varphi = (x^4 + 4tx)\varphi .$$

The verification of these properties is left to the reader.

The really amazing property is that $V(x, t)$ is a solution of the Korteweg-deVries equation

$$V_t = 3VV_x - \frac{1}{2}V_{xxx}$$

and one has

$$V(x, t) = -2 \frac{d^2}{dx^2} \log \theta_2(x, t) .$$

Notice, too, that $\theta' = 4\theta_2$.

This potential appears already in Bargmann's work on reflectionless potentials [19].

Example 2. We have checked that any rational solution of the Korteweg-deVries equation [20, 21] gives an instance where a) holds, i.e., a differential operator in λ exists.

A convenient way to obtain these potentials $V(x, t)$ is to use the recursion relation

$$\theta'_{k+1}\,\theta_{k-1} - \theta_{k+1}\,\theta'_{k-1} = (2k + 1)\theta_k^2$$

$$\theta_0(x) = 1 \qquad \theta_1(x) = x$$

and to set

$$V_k(x) = -2\frac{d^2}{dx^2}\log\theta_k .$$

This recursion relation was proposed by Adler and Moser [22, 23] and it is intimately related to the Darboux-Backlund-Crum method of going from one "solvable" Schroedinger problem to another such.

These potentials solve Lax, Gelfand-Dikki [24, 25] type equations of the form

$$V_{t_i} = [((-L)^{i+1/2})_+, L]$$

with $L = -d^2/dx^2 + V$, and the subindex plus denoting the differential operator part of a pseudodifferential operator.

Example 3. If $V(x)$ does not vanish at infinity then there are other instances of a), like the Airy's equation corresponding to a uniform field $V(x) = x$. For simplicity we consider only V that decay at infinity.

Example 4. Even under the restriction just mentioned we have not exhausted all the examples.

Consider the potential

$$V(x, t) = \frac{15x^4 - 18tx^2 - t^2}{4x^2(x^2 + t)^2} .$$

In this case we've

$$\varphi(x, \lambda) = \sqrt{x}\, J_2(\lambda x) - \frac{2t}{\lambda(t + x^2)} J_1(\lambda x)$$

and one can check that

$$[(-\frac{\partial^2}{\partial\lambda^2} + \frac{15}{4\lambda^2})^2 + 2t(-\frac{\partial^2}{\partial\lambda^2} - \frac{1}{4\lambda^2})]\varphi = (x^4 + 2tx^2)\varphi .$$

Notice that

$$V(x, t) = -\frac{9}{4}x^2 - 2\frac{d^2}{dx^2}\log(x^3 + xt)$$

and that

$$\theta' = 4(x^3 + xt) .$$

It is a fact that examples 1 and 4 are the only instances of a) with $M = 4$.

Before moving on to b), we notice the remarkable fact — which has been observed before by M. and Y. Sato — that the polynomials θ_k in Adler and Moser's hierarchy are exactly the characters of very special representations of $GL(n, C)$ or equivalently the Schur functions or the corresponding representation of S_n with $n = d(d + 1)/2$.

As to problem b) raised at the beginning of this section we notice that one can get good examples from the class of reflectionless potentials, as shown below.

Pick $c_i > 0$, $-k_i^2 < 0$, $i = 1, \ldots, n$ and put

$$M_{ij} = \delta_{ij} + c_j \frac{e^{(k_i+k_j)x}}{k_i + k_j} .$$

Now put

$$\tau(x) = \det M$$

and define

$$V(x) = 2\frac{d^2}{dx^2} \log \tau(x) .$$

A celebrated choice is

$$k_i = i, \quad c_j = 2j(-1)^{j-1} \prod_{i \neq j} \frac{i + j}{i - j}$$

leading to

$$V(x) = n(n + 1)\operatorname{sech}^2 x .$$

In this case the eigenfunctions are Legendre functions and one has well known 3 term recursion relations for these eigenfunctions. However any interesting deformation away from this potential destroys property b).

On the other hand we have found that picking an "isospectral manifold" of reflectionless potentials with great care, one obtains an entire manifold where property b) is preserved. Indeed the choice

$$k_i = i/2 \qquad i = 1, \ldots, n - 1 \qquad k_n = \frac{n^2 - n + 4}{4}$$

gives such a manifold. On it, the eigenfunctions satisfy a recursion relation with $M = (n^2 - n + 2)/2$, and everything can be written up fairly explicitly [26].

REFERENCES

[1] L. Shepp, Editor. Computed tomography. Proceedings of Symposia in Applied Mathematics, vol. 27. American Mathematical Society, Providence, Rhode Island, 1983.

[2] F. A. Grünbaum. The limited angle problem in tomography and some related mathematical problems. Internat. Colloq. Luminy (France), May 1982, North Holland.

[3] M. E. Davison and F. A. Grünbaum. 1979, Convolution algorithms for algorithms for arbitrary projection angles. IEEE Transactions on Nuclear Science, NS-26, 2670-2673.

[4] M. E. Davison and F. A. Grünbaum. 1981, Tomographic reconstruction with arbitrary directions. Communications on Pure and Applied Mathematics 34, 77-120.

[5] F. A. Grünbaum. A study of Fourier space methods for "limited angle" image reconstruction. Numerical Functional Analysis and Optimization 2(1), 31-42.

[6] D. Slepian. 1978, Prolate spheroidal wave functions. Fourier analysis and uncertainty. Bell System Tech. Journal 57, no. 5, 2371-1430.

[7] E. T. Whittaker. 1915, Proc. London Math. Soc. (2) 14, 260-268.

[8] E. L. Ince. 1922, On the connection between linear diff. systems and integral equations. Proc. of the Royal Society of Edinburgh (42), pp. 43-53.

[9] M. E. Davison. The ill conditioned nature of the limited angle tomography problem. SIAM J. Applied Math.

[10] M. E. Davison. 1981, A singular value decomposition for the Radon transform in n-dimensional euclidean space. Numer. Funct. Anal. and Optimiz. 3(3), 321-340.

[11] A. Papoulis. A new algorithm in spectral analysis and band limited extrapolation. IEEE Trans. on Circuits and Systems, vol. CAS-22, no. 9, 1975, 735-742.

[12] R. W. Gerchberg and W. O. Saxton. A practical algorithm for the determination of phase from image and diffraction plane pictures, Optik 35, no. 2, 1972, 237-246.

[13] S. Kacmarz. Bull. Int. Acad. Pol. Sci. Lett. Cl. Sci. Math. Nat. Ser. A, 1937, 355-357.

[14] D. C. Youla. Generalized image restoration by the method of alternating orthogonal projections. Image Science Mathematics Symposium, Monterey, California, 1976.

[15] F. A. Grünbaum. Limited Angle Reconstruction Problems in x-ray and NMR Tomography. Proceedings International Workshop on Physics and Engineering in Medical Imaging, IEEE, March 1982.

[16] F. A. Grünbaum. The Slepian-Landau-Pollak phenomenon for the radial part of the Laplacian. PAM-52, September 1981.

[17] F. A. Grünbaum. A new property of reproducing kernels for classical orthogonal polynomials. J. Math. Anal. Applic., to appear.

[18] J. Duistermaat and F. A. Grünbaum. In preparation, Differential equations in the eigenvalue parameter.

[19] V. Bargmann. On the connection between phase shifts and scattering potential. Rev. Mod. Phys. 21, 1949, 489-493.

[20] H. Airault, H. McKean, and J. Moser. 1977, Rational and elliptic solutions of the Korteweg-deVries equation and a related many body problem. Communications in Pure and Applied Math. (30), 95-148.

[21] D. Chudnovski and G. Chudnovski. 1977, Pole expansions for nonlinear partial differential equations. Nuovo Cimento, 40B, 339-353.

[22] M. Adler and J. Moser. 1978, On a class of polynomials connected with the Korteweg-deVries equation. Communications in Mathematical physics (61), 1-30.

[23] M. Ablowitz and H. Airault. 1981, Perturbations finies et forme particuliere de certaines solutions de l'equation de Korteweg-deVries. C. R. Acad. Sci. Paris, t. 292, 279-281.

[24] P. Lax. 1968, Integrals of nonlinear equations of evolutions and solitary waves. Communications on Pure and Applied Mathematics 21, 467-490.

[25] I. Gelfand and Dikii. 1976, Fractional powers of operators and Hamiltonian systems. Funkts. Anal. Prilozhen, 10, 4, 13-39.

[26] F. A. Grünbaum. Recursion relations and a class of isospectral manifolds for Schroedinger's equation, to appear in Nonlinear Waves, E. Debnath, Editor.

UNIVERSITY OF CALIFORNIA
MATHEMATICS DEPARTMENT
BERKELEY, CALIFORNIA 94720

Part III. Developments in Mathematical Inverse Theory

SIAM-AMS PROCEEDINGS
Volume **14**
1984

AN INVERSE SPECTRAL PROBLEM IN THREE DIMENSIONS

Roger G. Newton[1]

ABSTRACT. A generalization of the Gel'fand-Levitan method for the solution of the inverse spectral (inspec) problem to $\mathbb{R}^3$ is reviewed.

The solution of the inverse Sturm-Liouville problem has a long history (see, for example, [1] and [2]) and was given its most complete and elegant form by Gel'fand and Levitan [3a,b]. The best-known generalization of a version of that problem to two dimensions was given by Mark Kac in a lecture entitled "Can one hear the shape of a drum?" [4]. In the present lecture I will outline a solution of another version in three dimensions. For earlier partial solutions to the same problem by different methods, see [5c,d,6,7a,d].

The equation under consideration is the Schrödinger equation

$$[V(x) - \Delta]y = k^2y \ .$$

If posed in $\mathbb{R}$ this is an ordinary differential equation and two linearly independent solution $\phi_1(k^2,x)$ and $\phi_2(k^2,x)$ can be uniquely defined by boundary conditions at the origin. Under certain conditions on the potential $V(x)$ a family of such solutions for various real values of k^2 is known to allow a generalized Fourier transformation on $L^2(\mathbb{R})$ (see, e.g. [2]),

$$f(x) = \int (\hat{f}(k^2),\ d\rho(k^2)\ \phi(k^2,x)) \ ,$$

$$\hat{f}(k^2) = \int_{-\infty}^{\infty} dx\ \bar{f}\ (x)\ \phi(k^2,x) \ ,$$

in which ϕ is a vector function whose two components are the two solutions ϕ_1 and ϕ_2, (,) denotes the inner product on the two-dimensional space, and ρ is the <u>spectral distribution function</u> (or simply <u>spectral function</u>). The inverse spectral (inspec) problem is to determine the function $V(x)$ if $\rho(k^2)$ is given. In higher dimensions the first difficulty is that it is not obvious how to define an analogous spectral function.

1980 Mathematics Subject Classification. 35J10, 35R30, 81F05.
[1]Supported in part by the National Science Foundation.

From the point of view of the Gel'fand-Levitan solution to the inspec problem the crucial property of $\phi(k^2,x)$ is that for fixed x it is an entire analytic function of k, of exponential type $|x|$. It is this property that leads to the Povsner-Levitan representation with its famous "triangular" kernel [1]. The first task in a generalization to $\mathbb{R}^3$ is therefore to produce a family of solutions of the Schrödinger equation with similar properties.

As a preliminary step we define a scattering solution $\psi(k,\theta,x)$ of the Schrödinger equation by the requirement that

$$\lim_{|x| \to \infty} |x| e^{-ik|x|} [\psi(k,\theta,x) - E(k,\theta,x)]$$

exists. Here and in the following, $E(k,\theta,x) = \exp(ik\theta \cdot x)$, θ is a unit vector in $\mathbb{R}^3$, $|\theta| = 1$, $x \in \mathbb{R}^3$, and $\theta \cdot x$ denotes the inner product in $\mathbb{R}^3$. Equivalently, θ may be regarded as a point on the unit sphere S^2. It is useful to introduce the function $\gamma(k,\theta,x) = \psi(k,\theta,x)E(-k,\theta,x)$ and, for each fixed x, to consider $\gamma(k)$ and $\psi(k)$ as functions from $\mathbb{R}$ to a suitable Banach space B of functions $S^2 \to \mathbb{C}$. Their dependence on x will for the time being be suppressed as it plays no role. B will be specified later. Under fairly weak conditions on $V(x)$, which are certainly satisfied if $\exists\, a,\varepsilon,C > 0$ such that for all $x \in \mathbb{R}^3$ $|V(x)| \leq C(a + |x|)^{-3-\varepsilon}$ and $|\nabla V(x)| \leq C(a + |x|)^{-2-\varepsilon}$, it is known [7c] that $\gamma(k)$ is the boundary value of an analytic function that is meromorphic in $\mathbb{C}^+$, the open upper half of the complex plane, and such that there

$$\lim_{|k| \to \infty} \|\gamma(k) - \hat{1}\| = 0$$

if $\hat{1}$ denotes the function $\hat{1}(\theta) \equiv 1$. Furthermore

$$\int_{-\infty}^{\infty} dk \, \| \gamma(k) - \hat{1} \|^2 < \infty ,$$

where $\| \cdot \|$ denotes the norm in B. (In fact the square integrability of $\gamma - \hat{1}$ holds pointwise on S^2, and γ is continuous in θ and k, except possibly at $k = 0$.)

The function $\psi(k)$ is well known [7g] to satisfy the relation

$$\psi(-k) = \omega(k)Q\psi(k) ,$$

where $\omega(k)$ is a family of isometric operators from B to B, and Q is the operator defined by $(Qf)(\theta) = f(-\theta)$ for $f \in B$. (ω^{-1} is also known as the S matrix.)

On the assumption that $V(x)$ satisfies the stated conditions and that in addition its Fourier transform $\tilde{V}$ is such that $\exists\, C, \mu > 0$ so that for all $k \in \mathbb{R}^3$, $|\tilde{V}(k)| \leq C(\mu + |k|^2)^{-1}$, it has been proved in [7c] that the operator-valued function $\omega(k) - \mathbb{1}$ (where $\mathbb{1}$ is the unit operator) is strongly square-integrable, i.e., $\exists$ C such that for all $f \in B$

$$\int_{-\infty}^{\infty} dk \, \| [\omega(k) - \mathbb{1}] f \|^2 \leq C \| f \|^2$$

if B is taken to be $L^p(S^2)$ with $p > 4$. The Banach space B will therefore be so chosen from now on.

Let us now pose the following generalized Riemann-Hilbert problem *H*: Find an operator-valued function F(k) with these properties:

a) For $k \in \mathbb{R}$, $F(k) - \mathbb{1}$ is strongly square-integrable in the sense defined above.

b) For $k \in \mathbb{R}$, F(k) satisfies the relation

$$F(-k) = \omega(k) Q F(k) Q \ .$$

c) F(k) is the boundary value of an analytic function that is meromorphic in $\mathbb{C}^+$, with simple poles at those points k where k^2 is an eigenvalue of the Schrödinger equation (i.e., for those imaginary values of k where $\psi(k)$ has simple poles), such that the range of the residue of F at a pole equals the span of the residues of $\psi(k,-\theta,x)$ at the corresponding pole as x varies, and such that in $\mathbb{C}^+$, $\lim_{|k| \to \infty} \| F(k) - \mathbb{1} \| = 0$.

The Hilbert problem *H* does not necessarily have a solution, but if it and an associated Hilbert problem have solutions (which is the generic case) the solution to *H* is unique and can be constructed by solving a Fredholm integral equation of the second kind with a compact kernel [7f]. Furthermore its solution F(k) has an inverse $J(k) = F^{-1}(k)$ that is holomorphic in $\mathbb{C}^+$ and tends to $\mathbb{1}$, and $J(k) - \mathbb{1}$ is strongly square-integrable. J is an appropriate three-dimensional generalization of the Jost function.

We now define

$$\phi(k) = J(k)\psi(k) \ .$$

It then follows from the properties of J and ψ that $\phi(k,\theta,x)$ is a solution of the Schrödinger equation, $\phi(k)$ is holomorphic in $\mathbb{C}^+$, and it satisfies the relation

$$\phi(-k) = Q\phi(k) \ .$$

Consequently $\phi(k)$ is an <u>entire</u> analytic function [7c]. Furthermore, from the asymptotic form of $\psi(k)$ it follows that $\phi(k)$ is of exponential type $|x|$, and on the real axis $\phi(k) - E(k)$ is weakly square integrable: For each $a \in B^\dagger$ (where $B^\dagger$ is the dual of B)

$$\int_{-\infty}^{\infty} dk \, |(a,[\phi(k) - E(k)])|^2 < \infty$$

if (,) denotes the natural inner product for two functions in B and $B^\dagger$.

The solution-family $\phi(k,\theta,x)$ has all the relevant properties of the solutions in $\mathbb{R}$ defined by boundary conditions at the origin. It had to be defined by such elaborate methods because no direct procedure of defining it by boundary conditions is known. We may now proceed to define the spectral function.

Under very weak hypotheses on the potential the scattering solutions $\psi(k,\theta,x)$ are known to permit a generalized Fourier transformation. Regarding $\psi(k,\cdot,x)$ as a vector in $L^2(S^2)$ and using (,) as the inner product there, we have for $f \in L^2(\mathbb{R}^3)$

$$f(x) = \int_{k^2>0} (\hat{f}(k), d\sigma(k^2)\psi(k,x)) + \sum_n a_n u_n(x)$$

$$\hat{f}(k) = \int d^3x\, \bar{f}(x)\psi(k,x), \quad k^2 > 0$$

$$a_n = \int d^3x\, u_n(x) f(x) ,$$

where u_n are the normalized eigenfunctions, and $d\sigma(k^2)/dk^2 = (2\pi)^{-3}\, \tfrac{1}{2}k$ regarded as a multiplicative operator on $L^2(S^2)$. This "completeness relation" may be translated into a relation in terms of $\phi(k,x)$ and then reads

$$f(x) = \int (\hat{f}(k), d\rho(k^2)\phi(k,x))$$

$$\hat{f}(k) = \int d^3x\, \bar{f}(x)\phi(k,x) ,$$

where the <u>spectral function</u> ρ is defined by

$$d\rho(k^2)/dk^2 = \begin{cases} (2\pi)^{-3}\, \tfrac{1}{2}k\, [J(k)J^*(k)]^{-1} , & k^2 > 0 , \\ \sum_n M_n\, \delta(k^2 + \kappa_n^2) , & k^2 < 0 . \end{cases}$$

In this expression J^* stands for the adjoint of J on $L^2(S^2)$, $-\kappa_n^2$ are the eigenvalues, and M_n is an operator on $L^2(S^2)$ such that if $u_n^{(i)}(x)$ are the normalized eigenfunctions with eigenvalue $-\kappa_n^2$ then

$$\sum_i u_n^{(i)}(x)\, u_n^{(i)}(y) = (\phi(i\kappa_n,y),\, M_n\, \phi(i\kappa_n,x)) .$$

The spectral function has now been defined. It would be desirable to have a more direct definition of ϕ in terms of boundary conditions; the definition of the spectral function would then have a simpler meaning. Nevertheless, all the quantities here introduced are natural generalizations of their counterparts in one dimension.

The inspec problem in $\mathbb{R}^3$ is to determine V(x) if the kernel $\rho(k^2,\theta,\theta')$ of the operator family $\rho(k^2)$ is given as a function on $\mathbb{R} \times S^2 \times S^2$. It may be anticipated that this problem will in general have no solution. Because V(x) is a function of three real variables and $\rho(k^2,\theta,\theta')$ depends on five real variables, functions ρ that are admissible as spectral functions can be expected to be subject to severe restrictions. As we shall see, this expectation is borne out by our results.

As a first step, the stated analytic and asymptotic properties of $\phi(k,\theta,x)$ allow us to deduce the Povsner-Levitan representation

$$\phi(k,\theta,x) = e^{ik\theta\cdot x} - \int_{-|x|}^{|x|} d\alpha\; q(\alpha,\theta,x)e^{ik\alpha}$$

from the Paley-Wiener theorem [7d]. Since $\phi(-k,-\theta,x) = \phi(k,\theta,x)$ it follows that $q(-\alpha,-\theta,x) = q(\alpha,\theta,x)$. The Fourier transform $q(\alpha)$ of $\phi(k) - E(k)$ is well defined in the sense that for every $\alpha \in B^{\dagger}$ the function $f(\alpha) = (a,q(\alpha))$ belongs to $L^2(\mathbb{R})$ and its support lies in $[-|x|,|x|]$.

We may also define, in some sense, the three-dimensional Fourier transform of $\phi(k,\theta,x)$ if we regard $k\theta \equiv K$ as a vector in $\mathbb{R}^3$:

$$h(x,y) = (2\pi)^{-3} \int d^3K e^{-iK\cdot y}[e^{iK\cdot x} - \phi(k,\theta,x)] .$$

However, h will in general be a <u>distribution</u>. The function $q(\alpha,\theta,x)$ then ought to be the <u>Radon transform</u> of $h(y,x)$:

$$q(\alpha,\theta,x) = \int d^3y\; \delta(\alpha - \theta\cdot y)h(x,y) .$$

The crucial question now is whether $\operatorname{supp} q(\alpha,\theta,x) \subset [-|x|,|x|]$ implies that for fixed x, $\operatorname{supp} h(x,y) \subset \{y\ |y| \leq |x|\}$. The necessary and sufficient condition [8] for this is that q meet the infinitely many moment conditions

$$\int d\theta\; Y_\ell^m(\theta) \int_{-|x|}^{|x|} d\alpha\; \alpha^n\, q(\alpha,\theta,x) = 0$$

for all $\ell > n$, $|m| \leq \ell$, where Y_ℓ^m are spherical harmonics.

The following theorem has been proved [7e]: If in addition to satisfying the previously stated conditions, $V(x)$ decreases <u>exponentially</u>, in the sense that $\exists\ \varepsilon > 0$ so that

$$\int d^3x\; |V(x)|e^{\varepsilon|x|} < \infty$$

then in general the analytic function ϕ is such that

$$\int d\theta\; Y_\ell^m(\theta)\; \phi^{(n)}(0,\theta,x) = 0$$

for all $|m| \leq \ell$, $\ell > n$, where $\phi^{(n)}$ denotes the nth derivative with respect to k. This result (exceptions to which I have not been able to rule out; see [7e], Secs. 4 and 5) is equivalent to the needed moment conditions for the Fourier transform q of ϕ - E. As a consequence we many conclude that ϕ has a three-dimensional Povsner-Levitan representation

$$\phi(k,\theta,x) = e^{ik\theta\cdot x} - \int_{|y|\leq|x|} d^3y\, h(x,y)e^{ik\theta\cdot y} .$$

It should be re-emphasized that in its dependence upon y, h(y,x) can generally be expected to be a distribution. Nonetheless, this representation is the needed generalization of the famous triangularity condition that leads to the Gel'fand-Levitan equation.

In order to derive the generalized GL equation we follow the method of Kay and Moses [5a,b] and Faddeev [6]. Let us consider the three dimensional Povsner-Levitan representation as a mapping Ω, $\phi(k,\theta) = \Omega E(k,\theta)$, from $L^2(\mathbb{R}^3)$ augmented by the span of a finite number of exponentials to L^2. (That is, we use it both for the continuous spectrum and for the eigenvalues, for which $k = i|\kappa_n|$.) Then Ω has the kernel

$$\Omega(x,y) = \delta(x - y) - h(x,y) .$$

Its right inverse is $(\Omega^{-1})_r = \mathbb{1} + r$, where the kernel $r(x,y)$ of r satisfies the Volterra equation

$$r(x,y) = h(x,y) + \int_{|z|\leq|x|} d^3z\, h(x,z)\, r(z,y) .$$

It therefore exists, is unique, and the nullspace of Ω contains the null vector only. Consequently, the left inverse $(\Omega^{-1})_\ell = \mathbb{1} + \ell$ exists and is unique, and the kernel of ℓ satisfies the equation

$$\ell(x,y) = h(x,y) + \int_{|z|\geq|y|} d^2z\, \ell(x,z)\, h(z,y) .$$

It follows that the support of $\ell(x,y)$ must be contained in the ball $|y| \leq |x|$, and the two inverses are equal. Thus for $|y| < |x|$ we have

$$[\Omega^{*-1} - \Omega](x,y) = h(x,y)$$

if $A^* = \tilde{\bar{A}}$.

The "completeness relation" for ϕ may be written symbolically in the shorthand form

$$\int(\phi,d\rho\; \phi) = \delta$$

while that for E (i.e., for Fourier transformation) it may be written in the analogous form

$$\int(E,d\rho_0\; E) = \delta .$$

Inserting the Povsner-Levitan representation in the first leads to

$$\int(E,d\rho\; E)\tilde{\Omega} = \bar{\Omega}^{-1} ,$$

and in the second, to

$$\int(E, d\rho_0\, E)\tilde{\Omega} = \tilde{\Omega}\ .$$

Subtracting the two gives

$$\int(E, d(\rho - \rho_0)\, E)\tilde{\Omega} = \bar{\Omega}^{-1} - \tilde{\Omega}\ .$$

In terms of integrals this equation reads explicitly, for $|x| > |y|$

$$\text{(A)} \qquad h(x,y) = g(x,y) - \int_{|z|<|x|} d^3z\, h(x,z)\, g(z,y)\ ,$$

in which $g(x,y)$ is given by

$$g(x,y) = \int(E(x),\ d(\rho - \rho_0)E(y))\ ,$$

or more explicitly,

$$g(x,y) = \int(E(k,x),\ d[\rho(k^2) - \rho_0(k^2)]\ E(k,y))\ ,$$

$$d\rho_0(k^2)/dk^2 = \begin{cases} (2\pi)^{-3}\ \tfrac{1}{2}k^2\ \mathbb{1}\ , & k^2 > 0\ , \\ 0\ , & k^2 < 0\ . \end{cases}$$

Equation (A) constitutes the generalized Gel'fand-Levitan equation. Its solution in L^2 is easily proved to be unique [7c] if it exists. However, for distributional solutions a uniqueness proof has not yet been given.

If $\rho(k^2)$ is given, the inspec problem is solved by solving the generalized Gel'fand-Levitan equation (A) for the unknown function or distribution $h(x,y)$. Once h is determined, the three-dimensional Povsner-Levitan representation leads to the solution ϕ of the Schrödinger equation that contains the underlying potential $V(x)$. This potential is, however, more directly related to h as follows.

The scattering solution ψ of the Schrödinger equation is well-known to solve the Lippmann-Schwinger integral equation (see, e.g. [7g])

$$\psi(k,\theta,x) = e^{ik\theta\cdot x} - \frac{1}{4\pi}\int d^3y\ \frac{e^{ik|x-y|}}{|x - y|}\ V(y)\ \psi(k,\theta,y)\ .$$

If we use the definition $\phi = J\psi$ and take the Fourier transform with respect to k of the resulting equation we arrive at the equation, for $-|x| < \alpha < |x|$,

$$q(\alpha,\theta,x) = \xi(\alpha,\theta,x) - \int d\theta'\zeta(\alpha - \theta'\cdot x,\theta,\theta')$$

$$- \frac{1}{4\pi}\int d^3y\ \frac{V(y)}{|x - y|}\ q(\alpha - |x-y|,\theta,y)\ ,$$

where in the last integral $-|y| < \alpha - |x-y| < |y|$ and

$$\zeta(\alpha,\theta,\theta') = (2\pi)^{-1}\int_{-\infty}^{\infty} dk\ j(k,\theta,\theta')e^{-ik\alpha}\ ,$$

$$j(k,\theta,\theta') = J(k,\theta,\theta') - \delta(\theta,\theta') ,$$

$$\xi(\alpha,\theta,x) = \frac{1}{4\pi} \int d^3y \frac{V(y)}{|x - y|} \delta(|x-y| + \theta \cdot y - \alpha) .$$

The function ξ vanishes for $\alpha < \theta \cdot x$, and for $\alpha > \theta \cdot x$

$$\lim_{\alpha \downarrow \theta \cdot x} \xi(\alpha,\theta,x) = \tfrac{1}{2} \int_0^\infty dr\, V(x - \theta r) .$$

The function $\zeta(\alpha,\theta,\theta')$ is defined in a mean-square sense for almost all α, θ and θ' and, as a function of α, it need not be continuous. However, $\int d\theta' \zeta(\alpha - \theta' \cdot x,\theta,\theta')$ is continuous as a function of α for almost all θ and all x. As a result, for all fixed x and almost all fixed θ, $q(\alpha,\theta,x)$ has a step-like discontinuity equal to that of $\xi(\alpha,\phi,x)$ at $\alpha = \theta \cdot x$:

$$q(\theta \cdot x+,\theta,x) - q(\theta \cdot x-,\theta,x) = \tfrac{1}{2} \int_0^\infty dr\, V(x - \theta r) .$$

This equation implies that the discontinuity is differentiable with respect to x in the θ-direction, and

$$\text{(B)} \qquad 2\theta \cdot \nabla[q(\theta \cdot x+,\theta,x) + q(\theta \cdot x-,\theta,x)] = V(x) .$$

The potential is therefore determined in a simple manner from the Radon transform q of the solution h of the generalized Gel'fand-Levitan equation.

Note that (B) has a remarkable property: The left-hand side appears to depend on θ, but the right-hand side does not. If for the given spectral function ρ an underlying potential exists then (B) is guaranteed to hold. But if the existence of a potential is not a priori known then the left-hand side of (B) will be independent of θ only in exceptional cases which I call miracles. Only if the miracle occurs does a potential function exist. The miracle thus serves to characterize, at least in part, those functions $\rho(k,\theta,\theta')$ that are admissible as spectral functions of a Schrödinger equation.

A version of eq. (A) directly for the function q has also been derived (see [7e] Sec. 4), but it is cumbersome and I will not quote it here. There are also other Gel'fand-Levitan equations. One is a linear equation satisfied by the kernel that transforms the solution of one Schrödinger equation into that of another, if both spectral functions are known. The other is a nonlinear equation for that kernel. Both of these equations have been generalized to three dimensions [7e], at least formally. If h is a distribution the non-linear equation may not make any sense.

If the potential V(x) is assumed to be central, $V(x) = f(|x|)$, then it is a straightforward exercise in the use of Legendre expansions to reduce the solution of the inspec problem in $\mathbb{R}^3$ to a sequence of solutions of such problems in $\mathbb{R}_+$. Each of these has been shown [7e] to be identical to the well-known solutions (see [9] and [7g], chapter 20) by means of the Gel'fand-Levitan equation for central potentials.

BIBLIOGRAPHY

1. K. Chadan and P.C. Sabatier, Inverse Problems in Quantum Scattering Theory, Springer Verlag, New York, 1977.

2. M.A. Naimark, Linear Differential Operators, F. Ungar Publ. Co., New York, 1967, vol. 2.

3a. I.M. Gel'fand and B.M. Levitan, "On the determination of a differential equation by its spectral function," Dokl. Akad, Nauk SSSR 77 (1951), 557-560 (Math. Reviews 13, 240);

3b. ________, "On the determination of a differential equation from its spectral function," Isvest. Akad. Nauk SSSR 15 (1951), 309-356 (Am. Math. Soc. Transl. 1, 253-304).

4. M. Kac, "Can one hear the shape of a drum?," Am. Math. Monthly 73, no. 4, part II (1966), 1-23.

5a. I. Kay and H. Moses, "The determination of the scattering potential from the spectral measure function, I. Continuous spectrum," Nuovo Cimento 2 (1955), 917-961;

5b. ________, "The determination of the scattering potential from the spectral measure function, II. Point eigenvalues and proper eigenfunctions," Nuovo Cimento 3 (1956), 66-84;

5c. ________, "The determination of the scattering potential from the spectral measure function, V. The Gel'fand-Levitan equation for the three-dimensional scattering problem," Nuovo Cimento 22 (1961), 689-705;

5d. ________, "A simple verification of the Gel'fand-Levitan equation for the three-dimensional scattering problem," Commun. Pure and Appl. Math. 14 (1961), 435-445.

6. L.D. Faddeev, "Inverse problem of quantum scattering theory II," Itogi Nauki i Tekhniki, Sov. Prob. Mat. 3 (1974), 93-180 (Transl. J. of Soviet Math. 5 (1976), 334-396).

7a. R.G. Newton, "The Gel'fand-Levitan method in the inverse scattering problem," in Scattering Theory in Mathematical Physics, D. Reidel Publ. Co., Dordrecht, 1974, pp. 193-225.

7b. ________, "The three-dimensional inverse scattering problem in quantum mechanics," invited lectures at 1974 Summer Seminar on Inverse Problems, Am. Math. Soc., U.C.L.A., August, 1974.

7c. ________, "Inverse scattering II, Three dimensions," J. Math. Phys. 21 (1980), 1698-1715; 22 (1981), 631; 23 (1982), 693.

7d. ________, "Inverse scattering III, Three dimensions, continued," J. Math. Phys. 22 (1981), 2191-2200; 23 (1982), 693.

7e. ________, "Inverse scattering IV, Three dimensions: Generalized Marchenko construction with bound states," J. Math. Phys. 23 (1982), 594-604.

7f. ________, "On a generalized Hilbert problem," J. Math. Phys. 23 (1982), 2257-2265.

7g. ________, Scattering Theory of Waves and Particles, Second ed., Springer Verlag, New York, 1982.

8. D. Ludwig, Radon Transform on Euclidean Space, Commun. Pure Appl. Math. 19 (1966), 49-81.

9. R. Jost and W. Kohn, "On the relation between phase shift energy levels and the potential," Kgl. Danske Videnskab. Selskab, mat.-fys. Medd. 27 (1953), no. 9, 1-19.

DEPARTMENT OF PHYSICS
INDIANA UNIVERSITY
BLOOMINGTON, IN 47405

SIAM-AMS PROCEEDINGS
Volume **14**
1984

ISOSPECTRAL PERIODIC POTENTIALS ON $\mathbb{R}^n$

Gregory Eskin, James Ralston and Eugene Trubowitz

This is a report on [1], [2]. We consider an n-dimensional vector lattice L in $\mathbb{R}^n$ and smooth potentials on q on $\mathbb{R}^n$, satisfying the periodicity condition

$$q(x + d) = q(x) , \qquad \forall d \in L .$$

For such potentials one has the eigenvalue problems, parametrized by $k \in \mathbb{R}^n$,

$$-\Delta u + qu = \lambda u$$

$$u(x + d) = e^{2\pi i k\cdot d}u(x) , \qquad \forall d \in L ,$$

and the eigenvalues, with multiplicities $\lambda_1(k) \leq \lambda_2(k) \leq \lambda_3(k) \leq \cdots$ which we denote by $\text{Spec}_k(-\Delta + q)$. The spectrum of the Schrodinger operator $-\Delta + q$ on $L^2(\mathbb{R}^n)$ is known to be purely continuous and equal to the set $\bigcup_k \text{Spec}_k$, [3], [4]. The operator $-\Delta + q$ is frequently used as the Hamiltonian of an electron in a pure crystal lattice.

We have been concerned with the periodic spectrum, $\text{Spec}_0(-\Delta + q)$, which we denote simply by $\text{Spec}(-\Delta + q)$. This is just the spectrum of $-\Delta + q$ on the torus $\mathbb{R}^n/L$. However, if

(a) q is real analytic

(b) L has the property: $|d| = |d'| \Rightarrow d = \pm d'$, $\forall d, d' \in L$,

then Spec determines Spec_k for all k. Thus, under conditions (a) and (b), assuming one knows Spec_k for all $k \in \mathbb{R}^n$ is equivalent to knowing Spec. Given a basis $v_1,\ldots,v_n$ for L, (b) is implied by the rational independence of $v_i \cdot v_j$ $i \leq j$. The condition (b) does not hold for $L = \mathbb{Z}^n$, but neither, strictly speaking, does Spec determine Spec_k in that case. We will assume that L satisfies (b) from here on.

The central question for us has been: what is the set $I(q)$ of potentials $\tilde{q}$ such that

$$\mathrm{Spec}(-\Delta + \tilde{q}) = \mathrm{Spec}(-\Delta + q) .$$

$I(q)$ certainly contains $q(\underline{+} x + a)$ for $a \in \mathbb{R}^n$. There are quite a few q for which it contains much more. Let $L' = \{\delta \in \mathbb{R}^n : \delta \cdot d \in \mathbb{Z}, \forall d \in L)$ and let $S = \{\delta \in L' : \exists d \in L$ with $\delta \cdot d = 1\}$. The elements of S are just the vectors of minimal length in each direction in the dual lattice L'. Given a smooth Q on $\mathbb{R}$ satisfying $Q(s + 1) = Q(s)$, and $\delta \in S$

$$\mathrm{Spec}_k(-\Delta + Q(\delta \cdot x))$$

is determined by $\mathrm{Spec}(-|\delta|^2 \frac{d^2}{ds^2} + Q(s))$. Generically in Q, the set of $\tilde{Q}$ with

$$\mathrm{Spec}\left(-|\delta|^2 \frac{d^2}{ds^2} + \tilde{Q}\right) = \mathrm{Spec}\left(-|\delta|^2 \frac{d^2}{ds^2} + Q\right)$$

is an infinite dimensional torus.

Given a smooth periodic q we may expand q in a Fourier series

$$q(x) = \sum_{\delta \in L'} a_\delta e^{2\pi i \delta \cdot x} ,$$

and form the partial potentials

$$Q_\delta(s) = \sum_{n=-\infty}^{\infty} a_{n\delta} e^{2\pi i n s} .$$

Assuming q has mean zero so that $a_0 = 0$, we have the decomposition

$$q(x) = \frac{1}{2} \sum_{\delta \in S} Q_\delta(\delta \cdot x) .$$

THEOREM 1: Under hypotheses (a) and (b) $\mathrm{Spec}(-\Delta + q)$ determines

$$\mathrm{Spec}_k\left(-|\delta|^2 \frac{d^2}{ds^2} + Q_\delta(s)\right)$$

for all $\delta \in S$, $k \in \mathbb{R}$.

Before going further we need to review the spectral theory of $-|\delta|^2 \frac{d^2}{ds^2} + Q_\delta$ a bit. To describe isospectral manifolds here one considers Spec_0 and $\mathrm{Spec}_{\frac{1}{2}}$. Consider the eigenvalues and normalized eigenfunctions for

$$- |\delta|^2 \frac{d^2\varphi}{ds^2} + Q_\delta \varphi = \lambda\varphi$$

with the boundary conditions $\varphi(s) = \varphi(s + 1)$, giving Spec_o, and $\varphi(s) = -\varphi(s + 1)$, giving $\mathrm{Spec}_{\frac{1}{2}}$. We label the eigenvalues (rather idiosyncratically) as

$$\mathrm{Spec}_o = \{\lambda_0 \leq \lambda_2^- \leq \lambda_2^+ \leq \lambda_4^- \leq \lambda_4^+ \cdots\}$$

$$\mathrm{Spec}_{\frac{1}{2}} = \{\lambda_1^- \leq \lambda_1^+ \leq \lambda_3^- \leq \lambda_3^+ \leq \quad \cdots\} .$$

One finds by Floquet theory that

$$\lambda_0 < \lambda_1^- \leq \lambda_1^+ < \lambda_2^- \leq \lambda_2^+ \cdots$$

and $\bigcup_k \mathrm{Spec}_k$ is the set of bands $[\lambda_{i-1}^+, \lambda_i^-]$ $i = 1,2,\ldots$. The fundamental result on periodic potentials on $\mathbb{R}$ is that $I(Q_\delta)$ is an "analytic" torus whose dimension equals the number of m for which $\lambda_m^- < \lambda_m^+$, i.e. the number of open gaps in the spectrum (see [5]). For potentials near the zero potential one shows by the inverse function theorem that the set of gap lengths $\{\lambda_m^+(Q) - \lambda_m^-(Q)$, $m = 1,\ldots\}$ can be prescribed arbitrarily and conversely assuming Q is mean zero, the whole spectrum is a analytic function of the gap lengths. Recently Garnett and Trubowitz have shown that this result is true globally: any sequence of gaps in ℓ^2 is the gap sequence of a mean zero potential in L^2 and the gap sequence determines the periodic and anti-periodic spectra analytically. This makes it possible to construct nice families of potentials: given a finite set I, let $T(\varepsilon)$ be the set of Q for which $\lambda_m^+ - \lambda_m^- = \varepsilon\gamma_m$, $m \in I$, and $\lambda_m^+ = \lambda_m$, $m \notin I$. Then the Q in $\{T(\varepsilon) : 0 \leq \varepsilon \leq 1\}$ can be analytically parametrized as $Q(\alpha,\varepsilon;s)$, $(\alpha,\varepsilon) \in \mathbb{R}^I/(\pi Z)^I \times [0,1]$. Here

$$Q(\alpha,0;s) \equiv 0 \qquad \text{and}$$

$$\frac{\partial Q}{\partial \varepsilon}(\alpha,0;s) = \sum_{m\in I} \gamma_m \cos(2\pi ms + 2\alpha_m)$$

Now we return to $\mathbb{R}^n$, or more specifically $\mathbb{R}^2$. One can show

THEOREM 2: Given $\delta \in S$ under hypotheses (a) and (b) $\mathrm{Spec}(-\Delta + q)$ determines

$$\Phi_{\delta,m}(q) = \int_\Gamma |\nabla q_\delta|^2 (\varphi_m^+(\delta \cdot x))^2 dx$$

when $\lambda_m^+ > \lambda_m^-$. Here Γ is a fundamental domain for L and, given $d \in L\backslash 0$, such that $\delta \cdot d = 0$

$$q_\delta(x) = \sum_{\{\beta \in L' : \beta \cdot d \neq 0\}} \frac{a_\beta e^{2\pi i \beta \cdot x}}{2\pi i \beta \cdot d}$$

The one-dimensional constructions discussed earlier yield nice manifolds on which to check the power of the invariants $\Phi_{m,\delta}$. Let $\{\delta_1, \delta_2\}$ be a basis for L, and let $M(\varepsilon)$ denote the set of potentials on $\mathbb{R}^2$ of the form

$$q(x) = \sum_{i=1}^{N} q_i(\delta_i \cdot x)$$

where $q_i(s) \in T_i(\varepsilon)$, $i = 3,\ldots,N$ and $q_1(s)$ and $q_2(s)$ are fixed potentials with a single open gap. All potentials in $M(\varepsilon)$ are real analytic, and it follows from Theorem 1 that given a potential q in $M(\varepsilon)$ any analytic potential $\tilde{q}$ isospectral to q has a translate in $M(\varepsilon)$. Note further that the cardinality of the invariants

$$\Phi_{i,m} \overset{\text{def}}{=} \Phi_{\delta_i,m} ,$$

where $i = 3,\ldots,N$, $m \in I_i = \{m : \lambda_m^- < \lambda_m^+$ for potentials in $T_i(\varepsilon)\}$, is exactly the dimension of $M(\varepsilon)$, when $\varepsilon > 0$. Thus to prove rigidity theorems it suffices to check the independence of $\{\Phi_{i,m};\ i = 3,\ldots,N,\ m \in I_i\}$ on $M(\varepsilon)$. This is fairly easy to do near $\varepsilon = 0$, and by analyticity in the parameters gives the following.

THEOREM 3: For all but a finite set of ε in $[0,1]$, there is a closed nowhere dense set Σ_ε in $M(\varepsilon)$ such that, given $q \in M(\varepsilon)\setminus\Sigma_\varepsilon$ the set of analytic $\tilde{q}$ with

$$\mathrm{Spec}(-\Delta + \tilde{q}) \quad \text{and} \quad (-\Delta + q)$$

is <u>finite</u> modulo translations.

If one uses an invariant ψ from the Korteweg-de Vries series together with the $\Phi_{i,m}$ and restricts the form of $M(\varepsilon)$ a bit, requiring that $T_N(\varepsilon)$ consist of potentials with a single gap and δ_N have special properties relative to $\delta_1, \delta_2, \ldots, \delta_{N-1}$, one can finally reduce $I(q)$ to the minimal set mentioned earlier.

THEOREM 4: Given $M(\varepsilon)$ of the form described above there is is an open set $\mathcal{O}$ in $\{M(\varepsilon) : 0 < \varepsilon < \varepsilon_0\}$ such that given $q \in \mathcal{O}$, the set of analytic $\tilde{q}$ with

$$\mathrm{Spec}(-\Delta + \tilde{q}) \quad \text{and} \quad (-\Delta + q)$$

is given by $\{q(\pm x + a),\ a \in \mathbb{R}^2\}$.

Theorem 1 (as well as Theorem 2) is proven via spectral invariants. As is so often the case, these invariants are derived from trace theorems. Let the fundamental solution for the initial value problem for the heat equation

$$u_t = \Delta u - qu \quad \text{on} \quad R^n$$

be $G(x,y,t)$. Then

$$\sum_{\lambda \in \mathrm{Spec}_k} e^{-\lambda t} = \sum_{d \in L} e^{-2\pi i k \cdot d} \int_{\Gamma} G(x + d,x,t)dx$$

Hence, if one knows Spec_k for all k, one knows

$$\int_{\Gamma} G(x + d,x,t)dx$$

for $t > 0$, $d \in L$. Theorems 1 and 2 are derived from the asymptotics of $\int_{\Gamma} G(x + Nd + e,x,t)dx$ as $N \to \infty$.

If one considered the "wave trace"

$$W(t) = \sum_{\lambda \in \mathrm{Spec}_k} e^{\pm i\sqrt{\lambda}\, t}$$

then from the singularities of the distribution W one could get the same information as is contained the asymptotics of $\int_{\Gamma} G(x + d,x,t)dx$ as $t \to 0$. However, for our purposes it was useful to have asymptotics uniform in t. Moreover, to carry out the asymptotics of $\int_{\Gamma} G(x + Nd + e,x,t)$ to the order needed for Theorems 1 and 2 one only needs $q \in C^6(R^n/L)$. Thus, if one is willing to assume

$$\mathrm{Spec}_k(-\Delta + q) = \mathrm{Spec}_k(-\Delta + \tilde{q})$$

for all $k \in R^n$, then Theorems 1-4 hold with analyticity of $\tilde{q}$ replaced by $\tilde{q} \in C^6$.

REFERENCES

[1] G. Eskin, J. Ralston and E. Trubowitz, "On isospectral periodic potentials in $\mathbb{R}^n$", to appear in CPAM.

[2] G. Eskin, J. Ralston and E. Trubowitz, "On isospectral periodic in $\mathbb{R}^n$, II", preprint.

[3] J. Keller and F. Odeh, "Partial differential equations with periodic coefficients and Bloch waves in crystals", J. Math. Phys. 5 (1964), 1499-1504.

[4] L. Thomas, "Time dependent approach to scattering from impurities in a crystal", Comm. Math. Phys. 33 (1973), 335-343.

[5] E. Trubowitz, "The inverse problem for periodic potentials" CPAM **30** (1977) 321-337.

GREGORY ESKIN
Department of Mathematics
University of California
Los Angeles, California 90024

JAMES RALSTON
Department of Mathematics
University of California
Los Angeles, California 90024

EUGENE TRUBOWITZ
Department of Mathematics
ETH-Zentrum
ZURICH, Switzerland

Current address:
Department of Mathematics
University of California
Los Angeles, CA 90024

SIAM-AMS PROCEEDINGS
Volume **14**
1984

SOME ASPECTS OF INVERSE PROBLEMS IN SEVERAL-DIMENSIONAL WAVE PROPAGATION

William W. Symes[1]

ABSTRACT. We introduce a simple inverse problem for the acoustic wave equation with constant sound velocity, in which the density in a half space is to be recovered from the boundary values of the response to a normally incident step-like plane wave. We formulate this problem as a functional equation involving a trace operator, which maps the coefficient (density) to the boundary values of the solution. We show that the boundary values are as smooth (in the L^2-sense) as the coefficient, provided the coefficient satisfies additional tangential smoothness requirements. We also show that the trace operator is either Hölder - or Lipshitz - continuous in the appropriate sense, depending on the amount of tangential smoothness required. The arguments can be extended to yield error bounds for regularized nonlinear least-squares versions of the inverse problem.

Transient acoustic inverse (coefficient identification) problems concern the linear acoustic wave equation

$$\left[\frac{1}{\rho c^2}\partial_t^2 - \nabla\cdot\frac{1}{\rho}\nabla\right]u = 0$$

in a domain $\Omega \times [0,T]$ with boundary $\partial\Omega \times [0,T]$, $\Omega \subset \mathbb{R}^n$, together with appropriate boundary and initial conditions. Physically, u represents the excess pressure field in a fluid confined to Ω, with density ρ and sound velocity c, in the absence of body forces.

---

[1] 1980 Mathematics Subject Classification 35R25.
Supported by the National Science Foundation under grant number MCS-80-02996-01, and by Office of Naval Research under contract number N00014-83-K-0051.

The data of the inverse problem is another, independent boundary condition, prescribed on a (possibly proper) subset $\Sigma \times [0,T] \subset \partial\Omega \times [0,T]$. This additional data is meant to model measurements of some aspect of the field at the boundary, not already determined by the physics. For instance, if Dirichlet boundary data are prescribed, i.e., the pressure is given at the boundary, then Neumann data might constitute the additional data on Σ, which would model measurement of the normal particle acceleration at Σ.

Physics and experience indicate that, once the physical boundary and initial conditions are prescribed, the additional data, viewed as the response of the fluid to the physical conditions, depends on the coefficients ρ and c, i.e., on the mechanical constitution of the fluid. Conversely, the additional data carries some information about the coefficients. The solution of the inverse problem is the identification of the coefficients ρ and c from the additional information.

Regarding the physical boundary and initial conditions as given, the additional information is the product of some boundary operator B acting on u, and in turn is a functional of the coefficients ρ and c. Thus

$$\mathcal{T}(\rho,c) = Bu|_{\Sigma \times [0,T]}$$

The inverse problems is to invert this functional relation, i.e., solve the above equation for ρ and c, given the right-hand side.

Of course, the description just given is really of a class of inverse problems, depending on the selection of physical boundary conditions, additional data, type of domain, etc. Also, similar problems may be posed in linear elasticity and electromagnetism.

Models for many important technologies include inverse problems of this type, perhaps the best-known example being the seismic reflection method in petroleum exploration. Inverse problems have attracted correspondingly much attention, and have generated a considerable literature. Much of this literature concerns various algorithms for the approximate solution of the above functional equation.

This paper concerns the properties of the operator $\mathcal{T}$ as a map between appropriate function spaces. The determination of domain and range metrics for which $\mathcal{T}$ has such properties as continuity, differentiability, bounded invertibility of the derivative, and so forth, is obviously crucial in the design and analysis of algorithms for inverse problems, yet we believe that the study of this subject has barely begun.

We restrict our attention in this paper to the Special case $c \equiv$ constant, so that the rays of geometric acoustics are straight lines. Although this case is unrepresentative of practical problems, we feel that several key technical difficulties are still present, which are more easily understood in the straight-ray case than in general. Also, this special class of multidimensional problems is most closely related to acoustics in one space dimension, for which the rays can always be straightened by a coordinate transformation, under very mild smoothness assumptions on the coefficients.

Also, for the sake of simplicity, we consider $\Omega = \mathbb{R}^n$, and restrict the variable coefficient ρ to be constant on one side of a hyperplane. We probe the inhomogeneity with a disturbance which is for $t < 0$ a (progressing) plane wave supported on the constant-coefficient side of the hyperplane, with a step-discontinuity at its plane wavefront. These restrictions mostly can be removed, at the cost of increasing the complexity of the estimates.

We study the dependence of the trace of u and its normal derivative at the hyperplane, on the coefficient ρ.

We begin by relating the smoothness of ρ to the smoothness of the traces. If ρ is smooth (infinitely differentiable), then so are the traces. In dimension $n > 1$, however, if ρ has only a finite order of smoothness, in the Sobolev (L^2-) sense, then the traces are generally less smooth. In another place [1], we give examples which show that the formal linearization, of the relation between ρ and the traces, is neither bounded nor boundedly invertible as a map between Sovolev spaces of the same order. These examples are based on geometric optics, and show clearly that the problem comes from grazing rays, i.e., rays parallel to the bounding hyperplane.

We show here that if ρ is required to have sufficient additional tangential somothness, then the traces have the same (isotropic) order of smoothness as ρ (Theorem 1). Depending on the degree of additional tangential smoothness, we also show that the traces depend either Hölder-continuously, with exponent 1/2, or Lipshitz-continuously, on ρ. (Theorems 2 and 3).

These results are based on the use of "sideways" energy estimates for the wave equation viewed as an evolution equation in the normal ("depth") variable. Such estimates were used by Rauch and Taylor [2] to prove energy decay for mixed problems in one space dimension with dissipative boundary conditions. Since the time-like Cauchy problem for the wave equation in more than one space dimension is no longer hyperbolic, the estimates must be modified. In fact, the traces of tangential derivatives must be estimated separately to recover this type of argument. The crucial point is that the additional tangential smoothness of the coefficients allows us to estimate (space-time) Sobolev norms of the tangential derivatives of u, which in turn bound (lower) Sobolev norms of their traces via the trace theorem for distributions (we follow a slightly different argument below).

Clearly a pivotal role is played in all this by the time-like Cauchy problem. This ill-posed problem can be regularized, for smooth coefficients, by micro-local restrictions on the solution, as is explained in our previous paper [3]. These results depend on the propagation of singularities according to geometric optics, which generally fails for coefficients as singular as those allowed in this paper (although the results of Beals and Reed [4] allow a partial extension to non-smooth coefficients). It remains to be seen whether the tangential smoothness required in this paper can be relaxed to microlocal smoothness of some sort.

To end this introduction, we mention that the arguments used here can be extended to show that the relation between ρ and the traces of u is twice differentiable, to give bounds on the derivative and on its inverse, and on the second derivative, in the presence of appropriate tangential smoothness restrictions. These results combine with the downward-continuation method, developed in [5] for a one-dimensional problem, to yield error bounds, i.e., continuous dependence of ρ

on trace data, for a nonlinear least-squares version of the inverse problem, with the tangential smoothness as a regularizing <u>a-priori</u> bound. Since the tangentially smooth spaces are <u>not</u> compactly included in the corresponding isotropic Sobolev spaces, this regularization strategy is strictly less stringent than so-called Tihonov regularization. These results will be presented elsewhere.

We shall write the Cartesian coordinates in $\mathbb{R}^n$ as (x,z), $x = (x_1,\ldots,x_{n-1})$. We use the following notations for derivatives:

$$\partial_t = \frac{\partial}{\partial t}\ ;\quad \partial_z = \frac{\partial}{\partial z}\ ;\quad \nabla_x = \left(\frac{\partial}{\partial x_1},\ldots,\frac{\partial}{\partial x_{n-1}}\right)$$

$$\nabla = \left(\frac{\partial}{\partial x_1},\ldots,\frac{\partial}{\partial x_{n-1}},\frac{\partial}{\partial z}\right);\quad D_k = -i\,\frac{\partial}{\partial x_k}\ ,\quad k = 1,\ldots,n-1.$$

We use the hat $\hat{}$ to denote the partial Fourier transform in $(x_1,\ldots,x_{n-1})$, with Fourier variables $\xi = (\xi_1,\ldots,\xi_{n-1})$. The ($L^2$-) Sobolev spaces $H^k = H^k(\mathbb{R}^n)$ have their usual norms, denoted $\|\ \|_k$. Constants are denoted by C, and change their meaning from equation to equation. They depend on the dimension n, and on other choices only as noted.

We shall make use of the spaces $\mathcal{H}_{(m,s)}$ in Hörmander's notation - see [6], section 2.5. The facts we need concerning these spaces appear in the following statement:

(i) for positive integer m and real s, the topology on $\mathcal{H}_{(m,s)}$ is defined by the norm

$$\|u\|^2_{(m,s)} = \sum_{j=0}^{m} \int_{\mathbb{R}} dz \int_{\mathbb{R}^{n-1}} d\xi (1+|\xi|^2)^{(s+m-j)} |\partial_z^j\, \hat{u}(\xi,z)|^2$$

(ii) for any multi-index α, $D^\alpha u \in \mathcal{H}_{(p,t)}$ if $u \in \mathcal{H}_{(m,s)}$ and $|\alpha| \le m+s-(p+t)$, $\alpha_n \le m-p$

(iii) if $m+s > \frac{n}{2}$ and $m > \frac{1}{2}$, then $\mathcal{H}_{(m,s)} \subset L^\infty(\mathbb{R}^n) \cap C^0(\mathbb{R}^n)$ continuously.

<u>Lemma</u> 1. Suppose $f \in \mathcal{S}(\mathbb{R}^n)$, $g_2 \in L^2(\mathbb{R}^n)$, $g_1 \in L^2(\mathbb{R}^n) \cap L^\infty(\mathbb{R}_z, L^2(\mathbb{R}^{n-1}_x))$. Then for any $s > \frac{n-1}{2}$, there exists $C = C(s,n) > 0$ so that

$$\left|\int_{\mathbb{R}} dz \int_{\mathbb{R}^{n-1}} dx\, f g_1 g_2\right| \le C\|f\|_{(0,s)} \|g_2\|_{L^2(\mathbb{R}^n)} \|g_1\|_{L^\infty(\mathbb{R}_z, L^2(\mathbb{R}^{n-1}_x))} .$$

Proof. Set $g = g_1 g_2 \in L^1(\mathbb{R}^n)$. Then

$$\begin{aligned}
\left|\int dz \int dx fg\right| &= \left|\int dz \int d\xi\, \hat{f}(\xi,z)\hat{g}(\xi,z)\right| \\
&\le \int dz \|\hat{f}(z)\|_{L^1(\mathbb{R}^{n-1}_\xi)} \|\hat{g}(z)\|_{L^\infty(\mathbb{R}^{n-1}_\xi)} \qquad (1)\\
&\le C_s \int dz \{\int d\xi (1+|\xi|^2)^s |\hat{f}(\xi,z)|^2\}^{1/2} \|g(z)\|_{L^1(\mathbb{R}^{n-1}_x)}
\end{aligned}$$

By Cauchy-Schwartz, this is

$$\le C_s \|f\|_{(0,s)} \{\int dz [\int dx |g_1(x,z)||g_2(x,z)|]^2\}^{1/2}$$

Now

$$\begin{aligned}
&\int dz [\int dx |g_1(x,z)||g_2(x,z)|]^2 \\
&\quad \le \int dz \{\int dx |g_1(x,z)|^2\}\{\int dx |g_2(x,z)|^2\} \\
&\quad \le \|g_1\|^2_{L^\infty(\mathbb{R}_z, L^2(\mathbb{R}^{n-1}_x))} \int dz \int dx |g_2(x,z)|^2 \\
&\quad = \|g_1\|^2_{L^\infty(\mathbb{R}_z, L^2(\mathbb{R}^{n-1}_x))} \|g_2\|^2_{L^2(\mathbb{R}^n)}
\end{aligned}$$

whence the conclusion follows immediately. q.e.d.

Corollary 1. If, in the notation of the Lemma, f does not depend on $x_1,\ldots,x_k$, then the conclusion holds for any $s = \frac{1}{2}(n-k-1)$.

Suppose $\rho \in C^\infty(\mathbb{R}^n)$, $\rho(x,z) = 1$ for $z \le z_0$, some $z_0 > 0$, and $\rho(x,z) > 0$ for all x,z. Let u be the (distribution) solution of

$$\frac{1}{\rho}\partial^2 u - \nabla \cdot \frac{1}{\rho}\nabla u = 0 \qquad (2)$$

satisfying the initial condition

$$u(x,z,t) = H(t-z), \quad t < 0$$

(here H is the Heavyside unit step function).

It is standard that u is smooth in the forward light cone $\mathcal{C} = \{(x,z,t) : t > z\}$, and on the boundary $z = t$ undergoes a jump determined by the progressing wave expansion (geometric optics):

$$\lim_{\zeta\to 0^+} [u(x,z,z+\zeta) - u(x,z,z-\zeta)] = \rho^{1/2}(x,z) \qquad (3)$$

Furthermore u vanishes identically in the complement of $\bar{\mathcal{C}}$ (see for instance [7], Ch. VI, for statements which imply these).

Now u can also be viewed as the solution in $\mathcal{C}$ of the wave equation (2) satisfying the initial condition (3) for $t < 0$ (or $z \leq z_0$) and the characteristic initial condition

$$\lim_{\zeta \to 0^+} u(x,z,z+\zeta) = :\bar{u}(x,z) = \rho^{1/2}(x,z) \tag{4}$$

for $z > z_0$.

We shall be interested in the trace of u on the set $\Sigma \times [0,T]$, where $\Sigma \subset \mathbb{R}_x^{n-1} \cong \{(x,z) : z = 0\}$ is compact. Because of the finite-domain-of-dependence property of the wave equation, the values of ρ for large $|x| + |z|$ are irrelevant. Therefore we assume that $\log \rho \in C_0^\infty(\mathbb{R}^n)$.

We shall measure the size of $\log \rho$ in the $(1, s+1)$ norm for some choice of $s > \frac{n-1}{2}$, for which we adopt the concise notation

$$\|\varphi\|_{(1,s+1)} =: \|\varphi\|_c .$$

Note that from property (iii) above follows the estimate, for any $(x,z) \in \mathbb{R}^n$:

$$\exp[-C_s\|\log \rho\|_c] \leq \rho(x,z) \leq \exp[C_s\|\log \rho\|_c] \tag{5}$$

Lemma 2. The reflected energy

$$\|u(t)\|_E^2 := \int_{-\infty}^t dz \int_{\mathbb{R}^{n-1}} dx \, \frac{1}{\rho}(|\partial_t u|^2 + |\nabla u|^2)(x,z,t)$$

satisfies

$$\|u(t)\|_E \leq \|\log \rho\|_1 .$$

Proof. Suppose w is smooth in $\mathcal{C}$, has compact support in (x,z) for each t, and satisfies the inhomogeneous wave equation

$$\frac{1}{\rho} \partial_t^2 w - \nabla \cdot \frac{1}{\rho} \nabla w = f$$

in $\mathcal{C}$. Multiply both sides by $\partial_t w$, integrate over $\mathcal{C} \cap \{t \leq t_0\}$, and integrate by parts to obtain the basic energy identity

$$\|w(t_0)\|_E^2 = \int_{-\infty}^{t_0} dz \int dx \, \frac{1}{\rho}|\nabla \bar{w}|^2$$
$$+ 2\int_{-\infty}^{t_0} \int_{-\infty}^{t} dz \int dx \, f \, \partial_t w \tag{6}$$

where again $\bar{w}(x,z) := \lim_{\zeta \to 0^+} w(x,z,z+\zeta)$.

For $w = u$, $f \equiv 0$ and $\bar{w} = \rho^{1/2}$ so

$$\|u(t_0)\|_E^2 = \|\nabla \log \rho\|^2_{L^2(\mathbb{R}^{n-1}_x \times (-\infty, t_0])} \leq \|\log \rho\|_1 \qquad \text{q.e.d.}$$

For $z \leq T$ define

$$Q_T(z) := \frac{1}{2} \int_z^T dt \int dx \, [|\partial_t u|^2 + |\nabla u|^2]$$

Remark. Q_T is a sort of vertical energy form. Note that $u = \text{const.}$ for $z < t < -z$, $z < 0$, by the domain-of-dependence property, so $Q_T(z) \equiv 0$ for $z < -T$. Obviously $Q_T(z) = 0$ for $z > T$.

Lemma 3. For $j = 1, \ldots, n-1$, $0 \leq t \leq T$,

$$\|D_j u(t)\|_E \leq C \|\log \rho\|_c \sup_Z Q_T(z)^{1/2}$$

where $C = C(n,s,T)$.

Proof. We may rewrite the wave equation in the form

$$[\partial_t^2 - \nabla^2 + \nabla \log \rho \cdot \nabla] u = 0$$

whence

$$[\partial_t^2 - \nabla^2 + \nabla \log \rho \cdot \nabla] D_j u = -\nabla D_j \log \rho \cdot \nabla u .$$

The identity (6) applied to $w = D_j u$ yields

$$\|D_j u(t_0)\|_E^2 = \int_{-\infty}^{t_0} dz \int dx \, \frac{1}{\rho} \, |\nabla D_j \rho^{1/2}|^2$$

$$+ -2 \int_0^{t_0} dt_1 \int_{-\infty}^{t_1} dz \int dx \, \frac{1}{\rho} (\nabla D_j \log \rho \cdot \nabla u) \partial_t D_j u$$

(Note that $\nabla u \equiv 0$ for $z < t \leq 0$). So

$$\|D_j u(t_0)\|_E^2 \leq \|\log \rho\|_c^2 \exp\{C \|\log \rho\|_c\}$$

$$+ \exp\{C\|\log \rho\|_c\} \int_0^{t_0} dt_1 \int_{-\infty}^{t_1} dz \int dx |\nabla D_j \log \rho| |\nabla u| |\partial_t D_j u| .$$

We apply Corollary 1 with

$$f = |\nabla D_j \log \rho| \in \mathscr{X}_{(0,s)}(\mathbb{R}^n)$$

$$g_1 = \chi |\nabla u|$$

$$g_2 = \chi |\partial_t D_j u|$$

where χ is the characteristic function of $\mathcal{C} \cap \{t \leq t_0\}$ and $n \to n+1$. We obtain

$$\int_0^{t_0} dt_1 \int_{-\infty}^{t_1} dz \int dx |\nabla D_j \log \rho| |\nabla u| |\partial_t D_j u|$$

$$\leq C\|\log \rho\|_c \Big[\int_0^{t_0} dt_1 \int_{-\infty}^{t_1} dz \int dx |\partial_t D_j u|^2\Big]^{1/2}$$

$$\times \sup_z \Big(\int dx \int^{t_0} dt |\nabla u|^2\Big)^{1/2}$$

$$\leq C\|\log \rho\|_c^2 \sup_z Q_T(z) + \frac{1}{2}\int_0^{t_0} \|D_j u\|_E^2 .$$

The conclusion follows from Gronwall's inequality. q.e.d.

Theorem 1. There exists a continuous function $K(a,b)$ of $a,b > 0$ for which

$$Q_T(z) \leq K(\|\log \rho\|_c, T)$$

(K also depends on n and s).

Proof: Calculate

$$Q_T'(z) = -\frac{1}{2}\int dx[|\partial_t u|^2 + |\partial_z u|^2 + |\nabla_x u|^2](x,z,z)$$

$$+ \int_z^T dt \int dx[\partial_t u \partial_t \partial_z u + \nabla_x u \cdot \nabla_x \partial_z u + \partial_z u \partial_z^2 u](x,z,t)$$

$$= -\frac{1}{2}\int dx[|\partial_t u|^2 + |\partial_z u|^2 + |\nabla_x u|^2](x,z,z)$$

$$+ \int dx\ \partial_t u\ \partial_z u(x,z,t)\Big|_{T=z}^{T}$$

$$- \int_z^T dt \int dx\{-\partial_t^2 u + \nabla^2 u\}\partial_z u(x,z,t)$$

$$+ 2\int_z^T dt \int dx\ \nabla_x u \cdot \nabla_x \partial_z u(x,z,t)$$

$$= \int dx\ \partial_t u \cdot \partial_z u(x,z,T) - \frac{1}{2}\int dx |\nabla \bar{u}|^2(x,z)$$

$$- \int_z^T dt \int dx(\nabla \log \rho \cdot \nabla u)\partial_z u(x,z,t)$$

$$+ 2\int_z^T dt \int dx\ \nabla_x u \cdot \nabla_x \partial_z u(x,z,t) .$$

Since $Q_T(z) = 0$ for $z > T$, integrating from z_0 to ∞ we obtain

$$Q_T(z_0) \leq \int_{z_0}^T dz \int dx |\partial_t u \partial_z u(x,z,T)|$$

$$+ \frac{1}{2}\int_{z_0}^T dz \int dx |\nabla \bar{u}(x,z)|^2$$

$$+ \int_{z_0}^T dz \int_z^T dt \int dx |\nabla \log \rho| |\nabla u| |\partial_z u|$$

$$+ 2\int_{z_0}^T dz \int_z^T dt \int dx |\nabla_x u| |\partial_z \nabla_x u|$$

$$= I_1 + I_2 + I_3 + I_4 .$$

We estimate each term separately:

$$I_1 \leq \frac{1}{2}\int_{z_0}^{T} dz \int dx\{|\partial_t u|^2 + |\partial_z u|^2\}(x,z,T)$$

$$\leq \frac{1}{2}\exp\{C\|\log\rho\|_c\}\|u(T)\|_E^2$$

$$\leq C_1\|\log\rho\|_c^2 \exp\{C_2\|\log\rho\|_c\}$$

Since $\bar{u} = \rho^{1/2}$

$$I_2 = \frac{1}{8}\int_{z_0}^{T} dz \int dx\ \rho^{-1}|\nabla\rho|^2$$

$$\leq C_1 \exp\{C_2\|\log\rho\|_c\}\int_{z_0}^{T} dz \int dx|\nabla \log\rho|^2$$

$$\leq C_1\|\log\rho\|_c^2 \exp\{C_2\|\log\rho\|_c\}.$$

An intermediate estimate (1) in the proof of Lemma 1 allows us to write

$$I_3 \leq \int_{z_0}^{T} dz\{\int d\xi(1+|\xi|^2)^s|(\nabla\log\rho)\hat{\ }(\xi,z)|^2\}^{1/2} Q_T(z).$$

From Lemma 3,

$$\sum_{j=1}^{n-1}\int_{z_0}^{T} dz\int_z^T dt\int dx|\partial_z D_j u|^2$$

$$\leq \exp\{C_1\|\log\rho\|_c\}\sum_{j=1}^{n-1}\int_{z_0}^{T} dt\|D_j u(t)\|_E^2$$

$$\leq C_2\exp\{C_1\|\log\rho\|_c\}\ \|\log\rho\|_c^2\ Q^*$$

where $Q^* = \sup_z Q_T(z)$. On the other hand, Lemma 2 implies

$$\int_{z_0}^{T} dz\int_z^T dt\int dx|\nabla_x u|^2 \leq \exp\{C_1\|\log\rho\|_c\}\int_{z_0}^{T} dt\|u(t)\|_E^2$$

$$\leq C_2\exp\{C_1\|\log\rho\|_c\}\|\log\rho\|_c^2.$$

So the Cauchy-Schwarz inequality implies

$$I_4 \leq C_1\exp\{C_2\|\log\rho\|_c\}\|\log\rho\|_c^2\sqrt{Q^*}.$$

Adding up,

$$Q_T(z_0) \leq C_1\exp\{C_2\|\log\rho\|_c\}\|\log\rho\|_c(1+\sqrt{Q^*})$$

$$+ \int_{z_0}^{T} dz\{\int d\xi(1+|\xi|^2)^s|\nabla\log\rho|\hat{\ }(\xi,z)|^2\}^{1/2} Q_T(z).$$

Gronwall's inequality implies

$$Q_T(z) \leq C_1\|\log\rho\|_c \exp\{C_2\|\log\rho\|_c\}(1+\sqrt{Q^*})E(z)$$

where

$$E(z_0) = \exp\left\lfloor\int_{z_0}^{T} dz\{\int d\xi(1+|\xi|^2)^s|(\nabla\log\rho)^\wedge(\xi,z)|^2\}^{1/2}\right\rfloor$$
$$\leq \exp(T-z_0)^{1/2}\{\int dz\int d\xi\,(1+|\xi|^2)^s|(\nabla\log\rho)^\wedge(\xi,z)|^2\}^{1/2}$$
$$\leq \exp C\|\log\rho\|_C.$$

Thus

$$Q_T(z) \leq K_1(\|\log\rho\|_C, T)(1+\sqrt{Q^*})$$

where K_1 is exponential-polynomial in its arguments.

Since Q_T is smooth and of compact support, $Q_T(z^*) = Q^*$ for some z^*. This yields a quadratic inequality for Q^*, which we solve to obtain the asserted bound. q.e.d.

Let $M > 0$ and define

$$\mathcal{M} = \{\rho \in C^\infty(\mathbb{R}^n) : \rho \equiv 1,\ z < z_0,\ \text{and}\ \|\log\rho\|_C < M\}.$$

In the following, we replace uniform bounds on functions of ρ by functions of M. Thus the various constants which appear depend additionally on M.

For $\rho_1, \rho_2 \in \mathcal{M}$ and corresponding plane-wave responses u_1, u_2, define

$$v = u_1 - u_2$$
$$R_T(z) = \int_z^T dt\int dx(|\partial_t v|^2 + |\nabla v|^2)(x,z,t).$$

<u>Lemma</u> 4. For $0 \leq t \leq T$,

$$\|v(t)\|_E \leq C_1\|\log\rho_2 - \log\rho_1\|_C.$$

<u>Proof</u>. We apply (6) with $w = v$. The wave equation for v is

$$(\partial_t^2 - \nabla^2 - \nabla\log\rho_1\cdot\nabla)v = \sigma\cdot\nabla u_2$$

where $\sigma = \nabla(\log\rho_2 - \log\rho_1)$. Also

$$\bar{v} = \rho_2^{1/2} - \rho_1^{1/2} = \frac{\rho_2-\rho_1}{\rho_2^{1/2}+\rho_1^{1/2}}$$

Thus

$$\|\bar{v}\|_{H^1} \leq C\|\rho_2-\rho_1\|_{H_1} \leq C\|\log\rho_2 - \log\rho_1\|_C$$

So

$$\|v(t_0)\|_E^2 = \int_{-\infty}^{t_0} dz\int dx\,\frac{1}{\rho}\,|\nabla\bar{v}|^2$$
$$+\ 2\int_{-\infty}^{t_0} dt\int_{-\infty}^{t} dz\int dx\,\frac{1}{\rho}(\sigma\cdot\nabla u_2)\partial_t v.$$

As in the proof of Lemma 3, we estimate the last term using Lemma 1 and the uniform bound on

$$\int_z^T dt \int dx \ |\nabla u|^2 (x,z,t)$$

provided by Theorem 1. q.e.d.

<u>Theorem</u> 2. There exists $C_j = C(s,M,n,T) > 0$, $j = 1,2$ so that

$$\|v(\cdot,0,\cdot)\|_{H^1(\mathbb{R}^{n-1} \times [0,T])} \leq C_1 \|\log \rho_2 - \log \rho_1\|_C^{1/2}$$

$$\|\partial_z v(\cdot,0,\cdot)\|_{L^2(\mathbb{R}^{n-1} \times [0,T])} \leq C_2 \|\log \rho_2 - \log \rho_1\|_C^{1/2}$$

<u>Proof</u>. By computations analogous to those in the proof of Theorem 1, we arrive at the estimate

$$\begin{aligned} R_T(z_0) \leq & \int_{z_0}^T dz \int dx |\partial_t v \partial_z v| (x,z,T) \\ & + \frac{1}{2} \int_{z_0}^T dz \int dx |\nabla \bar{v}(x,z)|^2 \\ & + \int_{z_0}^T dz \int_z^T dt \int dx |\nabla \log \rho_1| |\nabla v| |\partial_z v| \\ & + 2 \int_{z_0}^T dz \int_z^T dt \int dx |\nabla_x v| |\partial_z \nabla_x v| \\ & + \int_{z_0}^T dz \int_z^T dt \int dx |\sigma| |\nabla u_2| |\partial_z v| . \end{aligned}$$

We estimate the first three terms exactly as in the proof of Theorem 1. For the fourth term, we first estimate (using Lemma 3)

$$\begin{aligned} \int_{z_0}^T dz \int dx |\partial_z \nabla_x v|^2 & \leq 2 \int_{z_0}^T dz \int dx (|\partial_z \nabla_x u_1|^2 + |\partial_z \nabla_x u_2|^2) \\ & \leq C_1 . \end{aligned}$$

So Lemma 4 and the Cauchy-Schwartz inequality imply that the fourth term is bounded by

$$C_2 \|\log \rho_1 - \log \rho_2\|_C .$$

To estimate the fifth term we use Lemma 1 again and the bound on

$$\int_z^T dt \int dx |\nabla u_2|^2$$

which follows from Theorem 1 as in the proof of Lemma 4. The upshot is

$$\begin{aligned} R_T(z_0) \leq C_1 d^2 + C_2 d + C_3 \int_{z_0}^T dz \{ [\int d\xi (1+|\xi|^2)^s |(\nabla \log \rho_1)^\wedge (\xi,z)|^2]^{\frac{1}{2}} \\ + 1 \} R_T(z) \end{aligned}$$

whence

$$R_T \leq C_4 d$$

where

$$d = \|\log \rho_1 - \log \rho_2\|_C \qquad \text{q.e.d.}$$

This argument fails to yield a Lipshitz bound because we lack an energy estimate on $\nabla_x v$ with which to bound the fourth term. In order to obtain such a bound, we must impose yet more tangential smoothness.

Let $\| \ \|_\ell = \| \ \|_{(1,s+2)}$ and set with $L > 0$

$$\mathcal{L} = \{\rho \in \mathcal{M} : \|\log \rho\|_\ell \leq L\}$$

Constants henceforth depend on L, along with everything else.

Set

$$S_T(z) = \frac{1}{2}\int_z^T dt \int dx[|\partial_t \nabla_x u|^2 + |\nabla\nabla_x u|^2](x,z,t)$$

$$S^* = \sup_z S_T(z)$$

<u>Lemma</u> 5. For $\rho \in \mathcal{L}$ we have

$$\|D_k D_j u\|_E(t) \leq C\sqrt{S^*}$$

$$k,j = 1,\ldots,n-1.$$

<u>Proof</u>. $D_k D_j u = w_{kj}$ satisfies and, $\bar{w}_{kj} = D_k D_j \rho^{1/2}$.

$$(\partial_t^2 - \nabla^2 + \nabla \log \rho \cdot \nabla) w_{kj} = \nabla D_k D_j \log \rho \cdot \nabla u$$

$$+ \nabla D_k \log \rho \, D_j \nabla u + \nabla D_j \log \rho \, D_u u^k$$

Now

$$\|\bar{w}_{kj}\|_{H^1} \leq C\|\log \rho\|_\ell$$

and we can estimate the r.h.s. of the wave equation for w, just as in the proof of Lemma 3. q.e.d.

<u>Lemma</u> 6. S^* is uniformly bounded on $\mathcal{L}$.

<u>Proof</u>. We proceed as in the proof of Theorem 2. With $w_k := D_k u$, $k = 1,\ldots,n-1$, we obtain

$$\begin{aligned}S_{T,k}(z_0) &:= \frac{1}{2}\int_{z_0}^{T} dt \int dx[|\partial_t w_k|^2 + |\nabla w_k|^2] \\ &\leq \int_{z_0}^{T} dz \int dx\ |\partial_t w_k \partial_z w_k|(x,z,t) \\ &+ \frac{1}{2}\int_{z_0}^{T} dz \int dx\ |\nabla \bar{w}_k|^2 \\ &+ \int_{z_0}^{T} dz \int_{z}^{T} dt \int dx |\nabla \log \rho||\nabla w_k||\partial_z w_k| \\ &+ 2\int_{z_0}^{T} dz \int_{z}^{T} dt \int dx |\nabla_x w_k||\partial_z \nabla_x w_k| \\ &+ \int_{z_0}^{T} dz \int_{z}^{T} dt \int dx |\nabla D_k \log \rho||\nabla u||\partial_z w_k| .\end{aligned}$$

We can now estimate the fourth term by $C\sqrt{S^*}$, and the rest is in the proof of Theorem 2. The boundedness of S^* on $\mathcal{L}$ now follows as in the proof of Theorem 1. q.e.d.

Lemma 7. For $k = 1,\ldots,n-1$, $0 \leq t \leq T$,

$$\|D_k v(t)\|_E^2 \leq C_1\ R^* + C_2 d^2 .$$

Proof. $D_k v$ solves

$$(\partial_t^2 - \nabla^2 + \nabla \log \rho_1 \cdot \nabla) D_k \cdot v = -\nabla D_k \log \rho_1 \cdot \nabla v$$
$$+ D_k \sigma \cdot \nabla u_2 + \sigma \cdot \nabla D_k u_2$$

(recall $\sigma = \nabla(\log \rho_2 - \log \rho_1)$) and

$$D_u \bar{v} = D_u[\rho_2^{1/2} - \rho_1^{1/2}] .$$

Now estimate the integrated term

$$\int_{z_0}^{T} dz \int_{z}^{T} dt \int dx\{-\nabla D_k \log \rho_1 \cdot \nabla v + D_k \sigma \cdot \nabla u_2 + \sigma \cdot \nabla D_k u_2\}\partial_t D_k v$$

in the energy identity by means of Lemma 1, Theorem 1 (second term), and Lemma 6 (third term) as in the proof of Lemma 3. q.e.d.

Theorem 3. There exist $C_1, C_2 > 0$ depending on L,T,s, and n, so that for $\rho_1, \rho_2 \in \mathcal{L}$,

$$\|(u_1 - u_2)(\cdot,0,\cdot)\|_{H^1(\mathbb{R}_x^{n-1} \times [0,T])}$$
$$\leq C_1 \|\log \rho_1 - \log \rho_2\|_{\ell}$$
$$\|u\partial_z(u_1 - u_2)(\cdot,0,\cdot)\|_{H^1(\mathbb{R}_x^{n-1} \times [0,T])}$$
$$\leq C_2 \|\log \rho_1 - \log \rho_2\|_{\ell} .$$

The proof is quite parallel to that of Theorem 1, and we omit it.

BIBLIOGRAPHY

1. W. Symes, "Inverse problems in several-dimensional wave propagation", Proc. Intl. Geophys. Rem. Sensing Symp., IEEE, 1983.

2. J. Rauch and M. Taylor, "Exponential decay of solutions to hyperbolic equations in bounded domains", Indiana J. Math., 24 (1974), pp.79-86.

3. W. Symes, "Continuation for solutions of wave equations: regularization of the time-like Cauchy problem", preprint (1983).

4. M. Beals and M. Reed, "Propagation of singularities for hyperbolic pseudo-differential operators with nonsmooth coefficients", Comm. Pure. Appl. Math., 35 (1982), pp.189-184.

5. W. Symes, "Impedance profile inversion via the first transport equation", J. Math. Anal. Appl., to appear (1983).

6. L. Hörmander, Linear Differential Differential Operators, Springer Verlag, New York-Berlin-Heidelberg, 1964.

7. R. Courant and D. Hilbert Methods of Mathematical Physics, Vol. II, J. Wiley/Interscience, New York, 1962.

Department of Mathematics
Michigan State University
East Lansing, MI 48823

SIAM-AMS PROCEEDINGS
Volume **14**
1984

IDENTIFICATION OF AN UNKNOWN CONDUCTIVITY BY MEANS OF MEASUREMENTS AT THE BOUNDARY

Robert V. Kohn*

and

Michael Vogelius*

ABSTRACT. We present a summary of results concerning the determination of an unknown conductivity by means of static measurements at the boundary. The main emphasis is on identifiability; we only briefly discuss the reconstruction problem. Some references are given to related work for time dependent problems.

1. INTRODUCTION. We study the following inverse problem: can one determine an unknown conductivity $\gamma(x)$ inside a body Ω, by means of static measurements at the boundary? A. P. Calderón raised this question in [2]; it may be seen as a natural extension of a problem analyzed by Cannon, Douglas and Jones in 1963 [3,4]. Despite some progress, many unsolved problems remain.

Our main goal here is to summarize what is known about identifiability; we do this in sections 2 and 3, which are based mostly on [18]. A few of the results presented - notably 2E and 3C - are previously unpublished. In section 4, we touch on the reconstruction problem, given finitely many measurements; and section 5 reviews the literature on some related problems. Potential applications include nondestructive testing and water resources management [14,13], but these will not be discussed here.

Throughout, Ω will be a bounded domain in $\mathbb{R}^n$, with unit normal ν along $\partial\Omega$. The unknown conductivity $\gamma(x)$ may take real scalar or matrix values, corresponding to an isotropic or anisotropic material. In the isotropic case

$$\gamma \in L^\infty(\Omega)\ , \quad \text{ess inf}\ \gamma > 0 \quad , \tag{1.1}$$

while in the anisotropic case

$$\gamma_{ij} = \gamma_{ji} \in L^\infty(\Omega)\ , \quad \lambda|\xi|^2 \leq (\gamma(x)\xi,\xi) \leq \Lambda|\xi|^2 \tag{1.2}$$

for all $x \in \Omega$ and $\xi \in \mathbb{R}^n$, with $\lambda > 0$. We consider solutions of

$$L_\gamma u = \nabla\cdot(\gamma(x)\nabla u) = 0 \quad ; \tag{1.3}$$

in the context of heat conduction u represents temperature and $\gamma\nabla u$ the heat

*Research supported in part by Mathematical Sciences Research Institute, Berkeley, CA (RVK) and ONR contract N00014-77-C-0623 (MV).

flux; in the context of direct current electrical conduction u represents voltage and $\gamma\nabla u$ is the vector of current flow. The natural things to measure at $\partial\Omega$ are the Dirichlet data $u|_{\partial\Omega}$ and the Neumann data $\gamma(x)\nabla u\cdot\nu|_{\partial\Omega}$. We denote by $P_\gamma : H^{1/2}(\partial\Omega) \to H^{-1/2}(\partial\Omega)$ the operator which associates the former to the latter,

$$P_\gamma\phi = \gamma\nabla u\cdot\nu|_{\partial\Omega} \text{ with } L_\gamma u = 0 \;,\; u|_{\partial\Omega} = \phi \;;$$

we shall say that "γ_1 and γ_2 give the same boundary measurements" if $P_{\gamma_1} = P_{\gamma_2}$.

Knowledge of P_γ yields the energy quadratic form

$$Q_\gamma(\phi) = \int_\Omega (\gamma\nabla u, \nabla u)\,dx = \int_{\partial\Omega} \phi\cdot P_\gamma\phi\,ds \;, \tag{1.4}$$

by Green's formula, where the rightmost integral represents the dual pairing of $H^{1/2}$ and $H^{-1/2}$. Conversely, Q_γ determines P_γ by the polarization identity; hence "γ_1 and γ_2 give the same boundary measurements" iff $Q_{\gamma_1}(\phi) = Q_{\gamma_2}(\phi)$ for each $\phi \in H^{1/2}(\partial\Omega)$. The following variational characterizations of Q_γ are well-known:

$$Q_\gamma(\phi) = \min_{\substack{w \in H^1(\Omega)\\ w=\phi \text{ on } \partial\Omega}} \int_\Omega (\gamma\nabla w, \nabla w)\,dx \tag{1.5}$$

$$-Q_\gamma(\phi) = \min_{\substack{\sigma\in L^2(\Omega;\mathbb{R}^n)\\ \nabla\cdot\sigma=0}} \left(\int_\Omega (\gamma^{-1}\sigma,\sigma)\,dx - 2\int_{\partial\Omega} \phi\sigma\cdot\nu\,ds \right) . \tag{1.6}$$

2. IDENTIFIABILITY - THE ISOTROPIC CASE. In one dimension, only the harmonic mean of γ can be detected by boundary measurement: it is an easy exercise to show that

2A. *For $\Omega = (a,b)$, γ_1 and γ_2 give the same boundary measurements iff they have the same harmonic mean.*

Fortunately, the situation is entirely different in dimension greater than one.

Cannon, Douglas, and Jones considered cylindrical domains, with γ constant along lines parallel to the axis, in 1963. They showed that such γ are identifiable:

2B [3]. *Suppose $\Omega = G \times (a,b)$, with G a bounded, $C^{2,\alpha}$ domain in $\mathbb{R}^{n-1}$, $\alpha > 0$. Then the elements of*

$$\Gamma_1 = \{\gamma \in C^{1,\alpha}(\bar{G}) \;:\; \inf \gamma > 0\}$$

can be distinguished by means of boundary measurements.

Their procedure for reconstructing γ is remarkably direct. Taking $(a,b) = (0,\pi)$ for simplicity, and writing $x' = (x_1,\ldots,x_{n-1})$, let $L_\gamma u = 0$ with

$$u = 0 \qquad \text{on} \quad G \times \{0,\pi\}$$

$$u = g(x')\sin x_n \quad \text{on} \quad \partial G \times (0,\pi) \ ,$$

where $g : \partial\Omega \to \mathbb{R}$ is any positive function. If

$$k(x') = \gamma(x') \left.\frac{\partial u}{\partial x_n}\right|_{x_n=0}$$

is measured, then

$$\gamma = k \cdot \exp(-w) \ ,$$

where $w(x')$ solves

$$\sum_{i,j=1}^{n-1} \frac{\partial}{\partial x_i}\left(k \frac{\partial w}{\partial x_j}\right) = k \qquad \text{on} \quad G$$

$$w = \ln g \quad \text{on} \quad \partial G$$

Thus $\gamma(x')$ is determined for all x', using a single choice of g.

The restriction that γ be independent of x_n is, of course, crucial to the preceeding analysis. One is not really finding γ "in the interior", since it is determined by its values along the boundary. When the dependence of γ is unrestricted, one naturally obtains information at the boundary more easily than in the interior. If everything is smooth, then γ is determined to infinite order at $\partial\Omega$.

2C [18]. *Suppose that* $\partial\Omega$ *is smooth, and that* $\gamma_1, \gamma_2 \in C^\infty(\bar{\Omega})$. *If* γ_1 *and* γ_2 *give the same boundary measurements, then*

(2.1) $$D^k\gamma_1 = D^k\gamma_2 \quad \text{on} \quad \partial\Omega$$

for all $k = (k_1,\ldots,k_n) \geq 0$, *where* $D^k = \left(\frac{\partial}{\partial x_1}\right)^{k_1} \cdots \left(\frac{\partial}{\partial x_n}\right)^{k_n}$.

The proof of 2C is local in character, but not constructive. For $x_0 \in \partial\Omega$ with $\nu_n(x_0) \neq 0$, consider the Dirichlet data

$$\phi_N(x) = N^{\frac{n}{2}-1} \prod_{j=1}^{n-1} \psi\,(N(x_j - x_{0,j}))$$

with corresponding solutions u_N^i,

$$L_{\gamma_i} u_N^i = 0 \ , \quad \left.u_N^i\right|_{\partial\Omega} = \phi_N \ , \quad i = 1,2 \ .$$

If $\psi \in C_0^\infty(\mathbb{R})$ has vanishing moments of order $\leq M-1$, then a version of "St. Venant's principle" provides that

$$|\nabla u_N^i(x)| \leq C\, N^{-M} \ , \quad N \to \infty$$

for $x \in \bar{\Omega}$ bounded away from x_0.

If (2.1) fails, then (relabeling if necessary)

$$\gamma_1(x) - \gamma_2(x) > C \text{ dist } (x,\partial\Omega)^{\ell}$$

with $\ell \geq 0$ and $C > 0$, in a neighborhood of some $x_0 \in \partial\Omega$; it now follows that $Q_{\gamma_1}(\phi_N) > Q_{\gamma_2}(\phi_N)$ provided $M > n\ell/2$ and N is sufficiently large, a contradiction to the fact that γ_1 and γ_2 give the same boundary measurements. Details may be found in [18]; see also 3C below.

An immediate corollary of 2C is this:

<u>2D : For a smoothly bounded domain</u> $\Omega \subset \mathbb{R}^n$, $n \geq 2$, <u>the elements of</u>

$$\Gamma_2 = \{\text{restrictions to } \Omega \text{ of positive, real analytic functions defined in a neighborhood of } \bar{\Omega}\}$$

<u>can be distinguished by boundary measurements.</u>

The analogue of 2D with "analytic" replaced by "piecewise analytic" is open. We are encouraged, however, by the following example.

<u>2E. Let</u> Ω <u>be the unit disc in</u> $\mathbb{R}^2$, <u>with polar coordinates</u> (r,θ) , <u>and denote by</u> Γ_3 <u>the set of conductivities</u>

$$\gamma(r,\theta) = \begin{cases} \gamma_0 & r < r_0 \\ 1 & r_0 \leq r < 1 \end{cases}$$

<u>with</u> $0 < r_0 < 1$ <u>and</u> γ_0 <u>a positive constant. Then the elements of</u> Γ_3 <u>can be distinguished by boundary measurements.</u>

To prove 2E , consider $\gamma,\tilde{\gamma} \in \Gamma_3$, with $\tilde{\gamma}$ corresponding to $\tilde{\gamma}_0$, $\tilde{r}_0$; we shall show that $Q_\gamma(\sin N\theta) \neq Q_{\tilde{\gamma}}(\sin N\theta)$ for all sufficiently large N , unless $\tilde{\gamma} = \gamma$. If $\tilde{r}_0 = r_0$ this follows instantly, since either $\tilde{\gamma} < \gamma$ or $\tilde{\gamma} > \gamma$ and consequently either $Q_{\tilde{\gamma}}(\sin N\theta) < Q_\gamma(\sin N\theta)$ or $Q_{\tilde{\gamma}}(\sin N\theta) > Q_\gamma(\sin N\theta)$, provided $\tilde{\gamma}_0 \neq \gamma_0$.

Relabeling if necessary, we assume that $\tilde{r}_0 < r_0$. For $N \geq 0$, let u_N solve

$$L_\gamma u_N = 0 \quad , \quad u_N\big|_{\partial\Omega} = \sin N\theta \quad ,$$

and notice that for $r < r_0$

$$u_N = c_N r^N \sin N\theta \quad , \quad c_N = \frac{2}{r_0^{2N}(1-\gamma_0)+(\gamma_0+1)} . \tag{2.2}$$

In case $\gamma_0 > 1$, (2.2) implies that

$$\int_{r<\tilde{r}_0} \tilde{\gamma}_0 |\nabla u_N|^2 dx < \int_{\tilde{r}_0<r<r_0} (\gamma_0-1)|\nabla u_N|^2 dx \tag{2.3}$$

for sufficiently large N, whence

$$\int_\Omega \tilde{\gamma}|\nabla u_N|^2 dx = \int_{r<\tilde{r}_0} \tilde{\gamma}_0|\nabla u_N|^2 dx + \int_{\tilde{r}_0<r<r_0} |\nabla u_N|^2 dx + \int_{r_0<r<1} |\nabla u_N|^2 dx$$

$$< \int_{\tilde{r}_0<r<r_0} \gamma_0|\nabla u_N|^2 dx + \int_{r_0<r<1} |\nabla u_N|^2 dx ,$$

so

(2.4) $$\int_\Omega \tilde{\gamma}|\nabla u_N|^2 dx < \int_\Omega \gamma|\nabla u_N|^2 dx = Q_\gamma(\sin N\theta) .$$

It follows, using (1.5), that $Q_{\tilde{\gamma}}(\sin N\theta) < Q_\gamma(\sin N\theta)$.

Next, suppose that $\gamma_0 < 1$. In this case, a similar argument gives

(2.5) $$\int_\Omega \tilde{\gamma}^{-1}|\sigma_N|^2 dx < \int_\Omega \gamma^{-1}|\sigma_N|^2 dx$$

with $\sigma_N = \gamma\nabla u_N$, and N sufficiently large. Using the dual variational principle (1.6), we conclude that $Q_{\tilde{\gamma}}(\sin N\theta) > Q_\gamma(\sin N\theta)$.

3. IDENTIFIABILITY - THE ANISOTROPIC CASE. In the anisotropic case, one can not expect to recover the full matrix γ_{ij} , as the following two examples demonstrate.

3A. Let $\Omega \subset \mathbb{R}^n$, $n \geq 1$, _and let_ γ _satisfy_ (1.2). _For any_ C^1 _diffeomorphism_ $\Phi : \Omega \to \Omega$ _with_

(3.1) $$\Phi(x) = x , \quad D\Phi(x) = I \quad \text{for all} \quad x \in \partial\Omega ,$$

let

$$\gamma^\Phi(\Phi(x)) = |\det(D\Phi(x))|^{-1} \cdot D\Phi(x)^t \cdot \gamma(x) \cdot D\Phi(x) .$$

Then all elements of

$$\Gamma_4 = \{\gamma^\Phi : \Phi \text{ satisfies } (3.1)\}$$

give the same boundary measurements.

We owe this remark to L. Tartar. If $L_\gamma u = 0$, then $L_{(\gamma^\Phi)} u^\Phi = 0$, with

$$u^\Phi(x) = u \circ \Phi^{-1}(x) ;$$

by (3.1), $u^\Phi = u$ on $\gamma^\Phi \cdot \nabla u^\Phi = \gamma \cdot \nabla u$ on $\partial\Omega$.

_3B_ [25]. _Let_ Ω _be the unit disc in_ $\mathbb{R}^2$, _with polar coordinates_ (r,θ) . _For any function_ $\alpha(r)$, _let_

$$\gamma^\alpha = \begin{pmatrix} \alpha\cos^2\theta + \alpha^{-1}\sin^2\theta & (\alpha-\alpha^{-1})\sin\theta\cdot\cos\theta \\ (\alpha-\alpha^{-1})\sin\theta\cdot\cos\theta & \alpha\sin^2\theta + \alpha^{-1}\cos^2\theta \end{pmatrix} .$$

Then all elements of

$$\Gamma_5 = \{\gamma^\alpha : \alpha \in L^\infty(0,1) , \text{ ess inf } \alpha > 0\}$$

give the same boundary measurements.

Indeed,

$$r \cdot L_{\gamma^\alpha} = \frac{\partial}{\partial r}\left(r\alpha(r)\frac{\partial}{\partial r}\right) + \frac{1}{r\alpha(r)}\frac{\partial^2}{\partial\theta^2} ;$$

note that when $\alpha \equiv 1$ this is just r times the Laplacian. For $N \in Z$, the solution of

$$L_{\gamma^\alpha} u_N = 0 , \quad u_N\big|_{\partial\Omega} = e^{iN\theta}$$

has the form $v(r)e^{iN\theta}$, with $v(1) = 1$ and

$$\frac{\partial}{\partial r}\left(r\alpha(r)\frac{\partial v}{\partial r}\right) - \frac{N^2}{r\alpha(r)} v = 0 .$$

This implies that

$$v(r) = c_1 \exp\left(|N|\int_1^r \frac{ds}{s\alpha(s)}\right) + c_2 \exp\left(-|N|\int_1^r \frac{ds}{s\alpha(s)}\right) ,$$

with $c_1 + c_2 = 1$. Since $v(r)e^{iN\theta} \in H^1(\Omega)$, c_2 must equal zero; hence the Neumann data associated to u_N is

$$\gamma^\alpha \nabla u_N \cdot \nu\big|_{r=1} = \alpha\frac{\partial u_N}{\partial r} = |N|e^{iN\theta} ,$$

regardless of the choice of α. The span of $\{e^{iN\theta}\}$ is dense in $H^{1/2}(\partial\Omega)$, so each $\gamma^\alpha \in \Gamma_5$ gives the same boundary measurements.

What *can* one detect, in the anisotropic case? The natural analogue of 2C is this: if (n-1) eigenvalues and eigenvectors of γ are known, the last eigenvalue can be distinguished by boundary measurements.

3C. Let $\gamma, \tilde\gamma$ be two symmetric, positive definite matrices with entries in $L^\infty(\Omega)$, and let $\{\lambda_i\},\{\tilde\lambda_i\}$ and $\{e_i\},\{\tilde e_i\}$ be the corresponding eigenvalues and eigenvectors. For $x_0 \in \partial\Omega$, let B be a neighborhood of x_0 relative to $\bar\Omega$, and suppose that

(3.2) $$\gamma,\tilde\gamma \in C^\infty(B) , \quad \text{and} \quad \partial\Omega \cap B \text{ is } C^\infty ;$$

(3.3) $$e_j = \tilde e_j , \quad \lambda_j = \tilde\lambda_j \quad \text{in } B , \quad 1 \le j \le n-1 ;$$

(3.4) $$e_n(x_0)\cdot\nu(x_0) \ne 0$$

(3.5) $$Q_\gamma(\phi) = Q_{\tilde\gamma}(\phi) \quad \text{for every} \quad \phi \in H^{1/2}(\partial\Omega)$$
$$\text{with} \quad \text{supp}\,\phi \subset B \cap \partial\Omega .$$

Then

(3.6) $$D^{\underline{k}}\lambda_n(x_0) = D^{\underline{k}}\tilde\lambda_n(x_0)$$

for any $\underline{k} = (k_1,\ldots,k_n) \ge 0$.

We sketch the proof. For a fixed $z \in \partial\Omega$ near x_0, let $\{\phi_N\}_{N=1}^{\infty}$ be a sequence of functions on $\partial\Omega$ such that

(3.7) $$\operatorname{supp}\phi_N \downarrow \{z\}$$

(3.8) $$\|\phi_N\|_{k,\partial\Omega\cap B} \leq C_k N^k \text{, for all } k \geq -M$$

(3.9) $$\|\phi_N\|_{0,\partial\Omega\cap B} = 1 \text{ ,}$$

where the norms are standard Sobolev norms based on L^2; the existence of such a sequence may be deduced from [18], for any fixed $M > 0$. If u_N solves

$$L_\gamma u_N = 0 \text{ in } \Omega \text{ , } u_N = \phi_N \text{ on } \partial\Omega \text{ ,}$$

then (3.7)-(3.9) imply

(3.10) $$\|u_N\|_{1,\Omega\setminus U} \leq CN^{-M}$$

for all neighborhoods U of z (c.f. Lemma 2 of [18]). Condition (3.4) gives in a sufficiently small neighborhood U of z

(3.11) $$\int_{U\cap\partial\Omega} |w|^2 ds \leq C \int_U |e_n\cdot\nabla w|^2 dx$$

for any $w \in H^1(\Omega)$ with $w = 0$ in $\Omega\setminus U$; using this and (3.7)-(3.9), one obtains

(3.12) $$\int_U \rho^\ell |e_n\cdot\nabla u_N|^2 dx \geq C_{\ell,\varepsilon} N^{-(n+1+\varepsilon)\ell} \text{ , } C_{\ell,\varepsilon} > 0$$

for any $\ell \geq 0$, $\varepsilon > 0$, with $\rho(x) = \operatorname{dist}(x,\partial\Omega)$, provided $M > (n+1)\ell/2$ (cf. Lemma 3 of [18]).

In order to prove (3.6) it is sufficient to verify that $\left(\frac{\partial}{\partial\nu}\right)^k\gamma = \left(\frac{\partial}{\partial\nu}\right)^k\tilde{\gamma}$ in a $\partial\Omega$ - neighborhood of x_0, for any $k \geq 0$. We prove this by contradiction, using (3.10) and (3.12). If it fails, we may assume (switching γ and $\tilde{\gamma}$ if necessary) that there exists a z near x_0 and a neighborhood $U \subseteq B$ of z such that

(3.13) $$\lambda_n(x) - \tilde{\lambda}_n(x) \geq C\rho(x)^\ell \text{ for } x \in U \text{ ,}$$

with $C > 0$, $\ell \geq 0$. Then

$$\int_\Omega (\gamma\nabla u_N,\nabla u_N)dx \geq \int_U (\gamma\nabla u_N,\nabla u_N)dx \geq \int_U (\tilde{\gamma}\nabla u_N,\nabla u_N)dx + C\int_U \rho^\ell |e_n\cdot\nabla u_N|^2 dx \text{ ,}$$

using (3.3) and (3.13). If $M > (n+1)\ell/2$, then (3.10) and (3.12) show that

$$C\int_U \rho^\ell |e_n\cdot\nabla u_N|^2 dx > \int_{\Omega\setminus U} (\tilde{\gamma}\nabla u_N,\nabla u_N)dx$$

for large N. Hence, using (1.5),

$$Q_\gamma(\phi_N) = \int_\Omega (\gamma\nabla u_N,\nabla u_N)dx > \int_\Omega (\tilde{\gamma}\nabla u_N,\nabla u_N)dx \geq Q_{\tilde{\gamma}}(\phi_N) \text{ ;}$$

This contradicts (3.5), and the proof is complete.

The preceding argument used (3.4) to justify (3.11). The following example suggests that (3.4) may be dispensable.

3D [4]. For $\Omega = (a,b) \times (c,d) \subset \mathbb{R}^2$, let Γ_6 denote the family of conductivities

$$\gamma(x) = \begin{pmatrix} 1 & 0 \\ 0 & \alpha(x_2) \end{pmatrix},$$

for $\alpha \in C^1$ with $\inf \alpha > 0$. Then the elements of Γ_6 can be distinguished by boundary measurements. More specifically, if $\gamma_i \in \Gamma_6$, $i = 1,2$ and

$$\gamma_1 \nabla u^1 \cdot \nu = \gamma_2 \nabla u^2 \cdot \nu \quad \text{at} \quad x_1 = b$$

where u^i is the solution to

$$L_{\gamma_i} u^i = 0 \quad \text{in} \quad \Omega$$
$$u^i = 0 \quad \text{at} \quad x_1 = a,b$$
$$u^i = \sin\left(\pi \frac{x_1 - a}{b-a}\right) \quad \text{at} \quad x_2 = c,d$$

then

$$\gamma_1 \equiv \gamma_2 .$$

4. RECONSTRUCTION. Setting aside the question of identifiability, how might one estimate γ in practice? A straightforward approach is the following: let V be a finite-dimensional subspace of $H^1(\Omega)$; for $w_j \in V$, $1 \leq j \leq m$, set $\phi_j = w_j|_{\partial\Omega} \in H^{1/2}(\partial\Omega)$, and measure $\gamma \frac{\partial u_j}{\partial \nu}$, where u_j solves

$$L_\gamma u_j = 0 \quad \text{in} \quad \Omega, \quad u_j|_{\partial\Omega} = \phi_j .$$

If G is a finite-parameter family of possible conductivities, and $\tilde{\gamma} \in G$, let $\tilde{u}_j \in V$ be the solution to the Galerkin equation

$$\int_\Omega (\tilde{\gamma} \nabla \tilde{u}_j, \nabla \phi) dx = 0 \quad \text{for all} \quad \phi \in V \cap \overset{\circ}{H}{}^1(\Omega)$$
$$\tilde{u}_j = \phi_j \quad \text{on} \quad \partial\Omega .$$

Now assume furthermore that V and G are selected so that $\tilde{\gamma} \frac{\partial w}{\partial \nu} \in H^{-1/2}(\partial\Omega)$ whenever $w \in V$ and $\tilde{\gamma} \in G$, and choose $\tilde{\gamma}$ to minimize

$$J(\tilde{\gamma}) = \sum_{j=1}^{m} \left\| \tilde{\gamma} \frac{\partial \tilde{u}_j}{\partial \nu} - \gamma \frac{\partial u_j}{\partial \nu} \right\|^2_{H^{-1/2}(\partial\Omega)} \tag{4.1}$$

among all $\tilde{\gamma} \in G$.

A variation of this method, and its finite-difference analogue, is studied by Falk in [11], for operators of the form

$$- \gamma u'' + c(x)u = f(x) \qquad 0 < x < 1$$

$$c,f \ \text{known}; \quad \gamma > 0 \ \text{constant}$$

in one dimension. He takes $G = \mathbb{R}_+$, and uses just one measurement (m=1) at one point on the boundary (x=0), so that the functional = 0 for some $\tilde{\gamma}$; and he estimates how $|\tilde{\gamma}-\gamma|$ depends on the choice of V.

The minimization of (4.1) has apparently not been studied in higher dimensions. There is, however, a large literature on parameter identification, much of which is relevant: see, for example, [10,19,22]. In other problems where identifiability and well-posedness are known, one can often prove the convergence of such a procedure [1]. And in some cases, even identifiability can be proved using an approximation algorithm [20].

Calderón takes a completely different approach to the reconstruction problem in [2], for the scalar case with $\delta = \|\gamma-1\|_{L^\infty}$ sufficiently small. He shows that the Fourier transform of γ (extended by zero off Ω) has the form

$$\hat{\gamma}(\xi) = f(\xi) + R(\xi) ,$$

where $f(\xi)$ can be determined by boundary measurements, and

$$|R(\xi)| \leq C\cdot\delta^2\cdot\exp(\pi\cdot|\xi|\cdot\mathrm{diam}(\Omega)) .$$

Thus boundary measurements suffice to approximate $\hat{\gamma}(\xi)$ well at low frequencies, if δ is small. For any fixed $\xi \in \mathbb{R}^n$, measuring $f(\xi)$ requires the Neumann data from just one solution of $L_\gamma u = 0$, corresponding to Dirichlet conditions $\exp[\pi(i\xi+\eta)\cdot x]$, with $\eta \in \mathbb{R}^n$, $\eta\cdot\xi = 0$, $|\eta| = |\xi|$.

5. RELATED WORK. The parabolic analogue of our problem is to determine $\gamma = \gamma(x,t)$, given overdetermined boundary data for solutions of

$$\frac{\partial u}{\partial t} - \nabla\cdot(\gamma\nabla u) = 0$$

on a space-time cylinder $\Omega \times (0,T)$. This problem has been studied extensively in space-dimension one, both as to the identifiability of γ and as to its numerical approximation: see [6] for γ = constant, [5,9,15,16] for $\gamma = \gamma(t)$, and [17,21,26] for $\gamma = \gamma(x)$. We know of no results in space dimension greater than one for spatially varying γ.

An interesting nonlinear analogue is obtained by letting γ depend on u, so that the equation becomes $\nabla\cdot(\gamma(x,u)\nabla u) = 0$. The case $\gamma = \gamma(u)$ has been studied in [7], and a related parabolic problem is treated in [8].

Many authors have studied the reconstruction of an unknown $\gamma(x)$, given knowledge of a single function u everywhere on Ω, satisfying $\nabla\cdot(\gamma\nabla u) = 0$. This problem is of particular interest for studying ground-water flow through porous rock. Identifiability is analyzed in [24], and the convergence of numerical schemes are studied in [23,12]. Applications and other numerical methods have been discussed in a dozen or so articles in Water Resources Research over the last ten years, of which [13,27] are examples.

BIBLIOGRAPHY

1. H. T. Banks and K. Kunisch, An approximation theory for nonlinear partial differential equations with applications to identification and control. SIAM J. Control & Opt., To appear.

2. A. P. Calderón, On an inverse boundary value problem. Seminar on Numerical Analysis and its Application to Continuum Physics. Soc. Brasileira de Matemática, Rio de Janeiro, 1980, pp. 65-73.

3. J. R. Cannon, J. Douglas, and B. F. Jones, Determination of the diffusivity of an isotropic medium. Int. J. Engng. Sci., 1, 1963, pp. 453-455.

4. J. R. Cannon and B. F. Jones, Determination of the diffusivity of an anisotropic medium. Int. J. Engng. Sci., 1, 1963, pp. 457-460.

5. J. R. Cannon, Determination of an unknown coefficient in a parabolic differential equation. Duke Math. J., 30, 1963, pp. 313-323.

6. J. R. Cannon, Determination of certain parameters in heat conduction problems. J. Math. Anal. & Appl., 8, 1964, pp. 188-201.

7. J. R. Cannon, Determination of the unknown coefficient k(u) in the equation $\nabla \cdot k(u) \nabla u = 0$ from overspecified boundary data. J. Math. Anal. & Appl., 18, 1967, pp. 112-114.

8. J. R. Cannon and P. Duchateau, Determining unknown coefficients in a nonlinear heat conduction problem. SIAM J. Appl. Math., 24, 1973, pp. 298-314.

9. J. Douglas and B. F. Jones, The Determination of a coefficient in a parabolic differential equation Part II. Numerical approximation. J. Math. Mech., 11, 1962, pp. 919-926.

10. P. Eykhoff, ed., Identification and System Parameter Estimation, parts 1 & 2, North-Holland, 1973.

11. R. S. Falk, Error estimates for the approximate identification of a constant coefficient from boundary flux data. Num. Funct. Anal. & Opt., 2, 1980, pp. 121-153.

12. R. S. Falk, Error estimates for the numerical identification of a variable coefficient. Math. Comp., To appear.

13. E. Frind and G. Pinder, Galkerkin solution of the inverse problem for aquifer transmissivity. Water Resour. Res., 9, 1973, pp. 1397-1410.

14. H. Fukue and K. Wada, Application of electric resistance probe method to non-destructive inspection. Mitsubishi Heavy Industries Technical Review, vol. 19, no. 2, 1982, pp. 83-94.

15. B. F. Jones, The determination of a coefficient in a parabolic differential equation Part I. Existence and uniqueness. J. Math. Mech., 11, 1962, pp. 907-918.

16. B. F. Jones, Various methods for finding unknown coefficients in parabolic differential equations. Comm. Pure & Appl. Math., 16, 1963, pp. 33-44.

17. S. Kitamura and S. Nakagiri, Identifiability of spatially-varying and constant parameters in distributed systems of parabolic type. SIAM J. Control & Opt., 15, 1977, pp. 785-802.

18. R. Kohn and M. Vogelius, Determining conductivity by boundary measurements. Submitted to Comm. Pure & Appl. Math.

19. C. S. Kubrusly, Distributed parameter system identification: a survey. Int. J. Control, 26, 1977, pp. 509-535.

20. J. L. Lions, Some aspects of modelling problems in distributed parameter systems. In Distributed Parameter Systems: Modelling and Identification, A. Ruberti (editor). Lecture Notes in Control and Information Sciences 1, Springer Verlag 1978, pp. 11-41.

21. A Pierce, Unique identification of eigenvalues and coefficients in a parabolic problem. SIAM J. Control & Opt., 17, 1979, pp. 494-499.

22. M. Polis and R. Goodson, Parameter identification in distributed systems: a synthesizing overview. Proc. IEEE, 64, 1976, pp. 45-61.

23. G. R. Richter, Numerical identification of a spatially varying diffusion coefficient. Math. Comp., 36, 1981, pp. 375-386.

24. G. R. Richter, An inverse problem for the steady state diffusion equation, SIAM J. Appl. Math., 41, 1981, pp. 210-221.

25. S. Spagnolo, Private communication.

26. T. Suzuki, Uniqueness and nonuniqueness in an inverse problem for the parabolic equation. J. Diff. Eqs., 47, 1983, pp. 296-316.

27. S. Yakowitz and L. Duckstein, Instability in aquifer identification: Theory and case studies. Water Resources Research, 16, 1980, pp. 1045-1064.

COURANT INSTITUTE OF MATHEMATICAL SCIENCES
NEW YORK UNIVERSITY
251 MERCER STREET
NEW YORK, NY 10012

DEPARTMENT OF MATHEMATICS AND
INSTITUTE FOR PHYSICAL SCIENCE AND TECHNOLOGY
UNIVERSITY OF MARYLAND
COLLEGE PARK, MD 20742

Part IV. Methods of Maximum Information Entropy

SIAM-AMS PROCEEDINGS
Volume **14**
1984

MAXIMUM-ENTROPY INVERSES IN PHYSICS

C. RAY SMITH, RAMARAO INGUVA AND R.L. MORGAN

Abstract. The image restoration problem for noiseless analog data is cast as an inverse problem and is then solved by a maximum-entropy approach. Several problems relating to time series, with and without noise, are addressed. Methods for improving the brute-stack procedure in seismology and for extending the Burg algorithm to problems with noise are presented.

1. **Introduction.** In the past, physicists have gone to great lengths in order to deal only with well-posed mathematical problems. While this practice has been essential to the discovery of many of the fundamental laws of nature, it has led to erroneous interpretations and treatments of at least one major area of theoretical physics, namely statistical mechanics. The development of statistical mechanics has followed a long, precipitous, and often retrogressive trail. Nevertheless, the formal and practical accomplishments in statistical physics have been considerable, while the conceptual ones have been comparatively meager – until recently.

The dogma which viewed the logical foundations of statistical mechanics as residing in dynamics succeeded in inflicting this area of physics with a myopia that has invaded other areas of physics. Jaynes [55, 56] was the first to provide convincing arguments that statistical mechanics should be identified as an ill-posed problem whose solution follows trivially from information-theoretic methods for determining generalized inverses. This philosophy, though not without earlier resistance, has had an unshakling effect, allowing the development of general methods for treating ill-posed problems in many areas of physics, as well as engineering, medicine, statistics, economics and search theory.

This and the accompanying articles by Jaynes, Skilling, and Shore should provide the reader with a good sample of the type of problems which are susceptible to solution by information-theoretic methods.

2. **A maximum-entropy inverse for analog images.** Maximum-entropy and Bayesian inverses for digitized images have been discussed in detail by Frieden [39, 40], Gull and Daniel [46], Jaynes [65, 68] and many others. We wish to outline here how the maximum-entropy generalized inverse for analog (or continuous) images and signals can be accomplished by using the maximum-entropy formalism for information-gathering regions, developed by Jaynes for nonequilibrium statistical mechanics [57, 63]. Our final result is a very concise and interesting expression for the generalized inverse; however, a major reason for presenting this result is that it allows us to concentrate on the Lagrange multiplier (a function) in our solution. The development of a rigorous description of the general form of this problem would entail a mathematical excursion which would likely obscure the

purpose of the present discussion. Accordingly, we concentrate on a problem which avoids the complications of the general problem.

We consider the imaging problem in which a scene provides the input signal v to a detector that responds with an output signal y. The input signal will depend upon several physical and geometrical variables characterizing both the scene and light source; we denote these variables by s. Similarly, the output y will depend upon a set of variables **u** prescribed by the design of the detector. Henceforth, we call v(s) the *scene* and y(**u**) the *image.*

It is assumed that the input and output of the detector are related as follows:

$$y(\mathbf{u}) = \int ds K(\mathbf{u},s) v(s). \tag{2.1}$$

We shall refer to the linear response function K(**u**,s) as the instrument function. In the problem under consideration K(**u**,s) and y(**u**) are known; the scene v(s) is the object of interest and must be determined by inverting Eq. (2.1). Moreover, it is assumed that K(**u**,s) is singular; this could occur because the detector maps each point of the scene onto a region of the detector, because the dimensionality of **u** is less than that of s, or because **u** is discrete and finite in number while s is continuous. As concrete examples of the above, a camera out of focus would correspond to the first case, and computerized x-ray tomography would fall under the other two cases. At any rate, it is assumed that Eq. (2.1) represents a generalized inverse problem.

A problem similar to the one posed above was solved by Jaynes in Sec.4.b of his Brandeis lectures [57] – a more complete treatment is given in Sec.D of his MIT lectures [63]. We shall modify our notation so that Eq. (2.1) acquires the same structure as the problem addressed by Jaynes. The change of notation is conceptually more delicate than is revealed here. Our treatment applies only to positive scalar signals and only to cases which yield to a frequency interpretation of probability theory.

It is assumed that v(s) is a non-negative luminous-flux density describing the light emanating from the scene; examples of such densities are radiance, radiant exitance and luminance. The total illuminance (used here as a nontechnical, generic term) is given by

$$P = \int ds v(s). \tag{2.2}$$

Normally, one can determine this quantity from Eq. (2.1) and the geometry of the problem. Next, we define a normalized density f(s) by

$$f(s) = v(s)/P \tag{2.3}$$

and a new instrument function by

$$U(\mathbf{u},s) = PK(\mathbf{u},s). \tag{2.4}$$

With these changes, Eq. (2.1) becomes

$$y(\mathbf{u}) = \int ds\, f(s) U(\mathbf{u},s). \tag{2.5}$$

The range of **u** is called the information-gathering region; we have data y(**u**) over this region.

If we maximize the entropy

$$H = -\int ds\, f(s) \log f(s) \tag{2.6}$$

subject to Eq. (2.5) as a constraint, we find

$$f(s) = Z^{-1} \exp[-\int d\mathbf{u} \lambda(\mathbf{u}) U(\mathbf{u},s)], \tag{2.7}$$

where the partition functional

$$Z[\lambda(\mathbf{u})] = \int ds \exp[-\int d\mathbf{u}\lambda(\mathbf{u})U(\mathbf{u},s)] \tag{2.8}$$

arises from the normalization of f(s). The Lagrange multiplier $\lambda(\mathbf{u})$, a function on the information-gathering region, is determined by requiring the "scene" in Eq. (2.7) to generate the image $y(\mathbf{u})$ as expressed by Eq. (2.5). This can be expressed concisely by

$$y(\mathbf{u}) = -\frac{\delta}{\delta\lambda(\mathbf{u})} \log(Z[\lambda(\mathbf{u})]). \tag{2.9}$$

The result in Eq. (2.7) is the maximum-entropy generalized inverse of Eq. (2.1). Of course, it is not the only solution of Eq. (2.1). Forceful and compelling agruments of its selection over all others abound in the literature [28, 66, 85, 97, 99, 103, 109] and need not be repeated here.

In the discretized version of maximum-entropy reconstruction, there is one Lagrange multiplier for each pixel of the image. Each such multiplier must be computed (usually by a lengthy iterative procedure) in order to insure that the predicted scene leads to the known image. The computational challenges and the number of Lagrange multipliers are so overwhelming that they tend to obscure the simple structure of the algorithm, as evidenced by Eq. (2.7). In several problems involving information-gathering regions, the Lagrange multiplier functions have a physical interpretation (e.g. inverse temperature as a function of position and time) and can be measured with special instruments (e.g. thermocouples) or can be calculated with the help of the data. Thus, we are led to inquire whether $\lambda(\mathbf{u})$ for a given imaging system can be identified with some quantity which can be measured by means of an auxiliary detector or can be extracted from the data. This line of investigation could lead to significant savings in computer time.

3. **Maximum-entropy methods in time-series analysis.** In an important class of problems, a detector generates output signals in response to input signals consisting of discrete time series of the form $\mathbf{Y} = (y_0, \ldots, y_N)$. The output signals may consist of different sequences of the time series or sets of numbers $A(y_i)$, $B(y_i, y_j), \ldots$ based on the time series. We discuss two problems involving real time series: one is related to the stacking problem in seismology; the other is of a general nature.

Brute stacking of seismic traces is deemed to be the most effective procedure for combining traces in seismic data processing. Next, we review this procedure. Suppose one has a collection of M common-depth-point traces:

$$\mathbf{Y}^1, \mathbf{Y}^2, \ldots, \mathbf{Y}^M, \tag{3.1}$$

where $\mathbf{Y}^r = (y_1{}^r, y_2{}^r, \ldots, y_N{}^r)$, $y_i{}^r = \tilde{y}_i{}^r + e_i{}^r$, and $e_i{}^r$ is the noise or error distorting the ith amplitude of the rth trace, $\tilde{y}_i{}^r$. The brute stack of these data is defined by

$$\langle y_i \rangle_0 = \frac{1}{M} \sum_{r=1}^{M} y_i{}^r, \; i = 1, 2, \ldots, N. \tag{3.2}$$

There is doubtlessly some redundancy in the data in the traces in Eq. (3.1); thus, it is hoped that brute stacking trades some information loss for noise cancellation. Nonetheless, the above procedure does not make maximal use of the information in the traces, nor does it allow effective use of prior information. We discuss a procedure which represents an improvement over the brute-stacking method.

Recently, Jaynes [65, 68] discussed the analysis of noiseless time series and gave a perspicuous derivation of the Burg algorithm [18-20] for the maximum-entropy spectral analysis of time series. For the purpose of making predictions, one is interested in the probability density p(Y). This quantity can be determined by maximizing the entropy

$$H = -\int dY\, p(Y) \log[p(Y)], \tag{3.3}$$

subject to constraints based on all data related to the time series. The available data for a case considered by Burg and Jaynes were the m autocorrelations defined by

$$R_k = \sum_{j=1}^{N-k} y_{j+k}\, y_j,\ k = 1,\ldots,m < N. \tag{3.4}$$

A determination of the maximum-entropy probability density based on Eq. (3.4) as constraints leads to the well-known Burg algorithm [18-20].

In the presence of noise, the above approach must be amended or replaced by another, as we discuss next. A very effective way for taking Gaussian noise into account in the maximum-entropy formalism was employed by Gull and Daniell [46] for image reconstruction. An approach similar in spirit to that of Gull and Daniell can be brought to bear on the stacking problem.

The noise in the amplitudes of each trace in Eq. (3.1) can be taken into account through the variable

$$A_r = \sum_i (\hat{y}_i^r - \langle y_i \rangle_0)^2 / (e_i^r)^2, \tag{3.5}$$

where the $(e_i^r)^2$ must be approximated from the data and $\hat{y}_i^r$ represents our estimate of $\tilde{y}_i^r$. Depending on the confidence in the data, one can assign a numerical estimate for A_r. Using Eq. (3.1), we can estimate the autocorrelations R_k^r for each trace. Using these as data to generate constraints, Eq. (3..5) as a constraint, and any other available constraints in the maximum-entropy formalism, we obtain a maximum-entropy estimate $P(\hat{Y}^r)$ for the probability density $P(\tilde{Y}^r)$ for the noiseless trace $\tilde{Y}^r = (\tilde{y}_1^r,\ldots,\tilde{y}_N^r)$.

For computational purposes, we must discretize the above results. In the discrete representation, we define statistical weights w_r as follows:

$$w_r = p(\hat{Y}^r),\ r = 1,\ldots,M. \tag{3.6}$$

The stacking of the common-depth-point traces is then accomplished by

$$\langle y_i \rangle = \sum_{r=1}^{M} w_r y_i^r. \tag{3.7}$$

We see that the weights 1/M in Eq. (3.2) are replaced by w_r in Eq. (3.7). Obviously, the brute stack is recovered automatically should the above procedure render the w_r equal. However, a portion of the information lost by the brute-stacking method will be preserved by the procedure just described. Studies using real and synthetic data [53] indicate that this new method of stacking is superior to the brute-stacking method. Admittedly, the new procedure is somewhat ad hoc – but it works. Formal justification of this procedure is a subject of current research. Below, we indicate a line of reasoning being pursued to substantiate this approach to time series analysis.

The rationale for the Gull and Daniell algorithm [46] was provided by Jaynes, using a Bayesian analysis of the problem [65, 68]. Suppose instead of Eq. (3.4) we have data

$$d_k = R_k + n_k, \tag{3.8}$$

where n_k is the noise contributing to the measurement of R_k. Provided the probability density of the noise (n_k) is known, one can use this density and Eq. (3.8) to determine the likelihood function for use in a Bayesian analysis of this problem. However, for the case of a time series, it is necessary to generalize the counting procedure Jaynes used for determining the prior distribution. This extension is necessary because a time series, unlike an image, will ordinarily consist of both positive and negative amplitudes. We do not give the details here of the requisite extension. Suffice it to note that the tensor product of the hypothesis (or sample) space used for the imaging problem and the Pauli spin space of quantum mechanics will provide a suitable hypothesis space for the problem under discussion.

Given the generalization just described, one can readily determine the prior distribution to be used in Bayes' theorem in order to arrive at the posterior distribution for the amplitudes of the time series. Finally, the posterior distribution is maximized to estimate the time series. If the only data are of the form given in Eq. (3.8), this method will fail to give an overall phase factor. Extensions of the above analysis to other problems are fairly straightforward. In case the noise (e_i^r) is Gaussian, this approach can be used to partially justify the procedure suggested above for stacking.

The material in this section is somewhat qualitative in nature. Its purpose is to give a few suggestions on a fundamental line of reasoning that can be used to analyze several problems receiving intense research effort.

4. **Literature survey.** At the end of this article there appears a functional bibliography on maximum-entropy methods in applied statistics. In order to make this bibliography compendious, we have selected out of all publications on each topic, the ones which are both inherently of interest and, further, reference many other works on the topic. Several key papers do not appear in the bibliography; however, we have attempted to insure that they are referenced in entries there. Accordingly, if one is interested in establishing priorities of major contributions to the subject, it will be necessary to consult the reference sections of the entries in our bibliography. With but a handful of exceptions, the papers, books, and reports selected for the bibliography exist in a readily accessible medium.

ACKNOWLEDGEMENTS. The sheer number of citations of his works indicates our considerable indebtedness to Professor Edwin T. Jaynes. Our friend and mentor has not only been a source of encouragement, but has also shared freely his ideas and his unpublished manuscripts.

The first author has profited from discussions with Dr. J.L. Johnson. The pleasant working environment at the Research Directorate of the U.S. Army Missile Laboratory, MICOM, made possible by Dr. J.S. Bennett, Dr. R.L. Hartman and Dr. G.M. Duthie, is gratefully acknowledged.

Part of this work was done while the second author was visiting Conoco, Inc. (December, 1982), and R. Inguva expresses his gratitude to Dr. R.H. Stolt for his hospitality during this visit.

BIBLIOGRAPHY

1. J.G. Ables, "Maximum entropy spectral analysis," Astron. Astrophys. Suppl. **15** (1974, 383-393.

2. J. Aczel and Z. Daroczy, *On Measures of Information and Their Characterizations,* Academic Press, New York (1975).

3. M. Agu, "Bayesian analysis of nonlinear fluctuating-measurement systems," Phys. Rev. A. **24** (1981), 2174-2179.

4. Y. Alhassid, N. Agmon and R.D. Levine, "An upper bound for the entropy and its applications to the maximal entropy principle," Chem. Phys. Lett. **53** (1978), 22-26.

5. D. Anastassiou and D.J. Sakrison, "A probability model for simple closed random curves," IEEE Trans. Info. Th. **IT-27** (1981), 376-381.

6. V. Ya. Anisimov and B.A. Sotskii, "On the relation between the coherence and entropy of radiation," Theor. Math. Phys. **14** (1974), 211-213.

7. G.S. Arnold and J.L. Kinsey, "Information theory for marginal distributions: application to energy disposal in an exothermic reaction," J. Chem. Phys. **67** (1977), 3530-3532.

8. R. Balian, "Random matrices and information theory," J. Phys. Soc. Japan **26** (1969), 30-34.

9. N. Bara and K. Murata, "Maximum entropy image reconstruction from projections," Optics Comm. **38** (1981), 91-95.

10. Y. Bard, "Estimation of state probabilities using the maximum entropy principle," IBM J. Res. Develop. **24** (1980), 563-569.

11. W.H. Barker, "Information theory and the optimal detection search," Op. Res. **25** (1977), 304-314.

12. B. Berliner and B. Lev, "On the use of the maximum entropy concept in insurance," Trans. 21st Int. Congress of Actuaries, Switzerland (1980), 47-61.

13. M. Bertero, C. De Mol and G.A. Viano, "On the problems of object restoration and image extrapolation in optics," J. Math. Phys. **20** (1979), 509-521.

14. I. Bialynicki-Birula and J. Mycielski, "Uncertainty relations for information entropy in wave mechanics," Commun. Math. Phys. **44** (1975), 129-132.

15. R.G. Bowers and A. McKerrell, "The information-theoretic statistical mechanics of a system in contact with a heat reservoir," Am. J. Phys. **46** (1978), 138-142.

16. T.M. Brown, "Information theory and the spectrum of isotropic turbulence," J. Phys. A. **15** (1982), 2285-2306.

17. R.K. Bryan and J. Skilling, "Deconvolution by maximum entropy, as illustrated by application to the jet of M87," Mon. Not. R. astr. Soc. **191** (1980), 69-79.

18. J.P. Burg, "Maximum entropy spectral analysis," Proc. 37th Meeting Soc. Explor. Geophys., 1967.

19. J.P. Burg, "The relationship between maximum entropy spectra and maximum likelihood," Geophys. **37** (1972), 375-376.

20. J.P. Burg, "Maximum entropy spectral analysis," Ph.D. Dissertation, Stanford University, Stanford, CA (1975).

21. J.P. Burg, D.G. Luenberger and D.L. Wenger, "Estimation of structured covariance matrices," Proc. IEEE **70** (1982), 963-974.

22. J. Campbell, *Grammatical Man,* Simon & Schuster, New York (1982).

23. G.J. Chaitin, "Information-theoretic computational complexity," IEEE Trans. Inform. Theory **IT-20** (1974), 10-15.

24. G.J. Chaitin, "Algorithmic information theory," IBM J. Res. Develop. **21** (1977), 350-359, 496.

25. G.J. Chaitin, "Goedel's theorem and information," Int. J. Theor. Phys. **22** (1982), 941-954.

26. C.H. Chen, *Nonlinear Maximum Entropy Spectral Analysis Methods for Signal Recognition,* Wiley, New York (1982).

27. D.G. Childers, ed., *Modern Spectrum Analysis,* IEEE Press, New York (1978).

28. R. Christensen, *Entropy Minimax Sourcebook,* Entropy Ltd., Lincoln, Mass. (1981), Vols. I-IV.

29. R.T. Cox, "Probability, frequency and reasonable expectation," Am. J. Phys. **14** (1946), 1-13.

30. R.T. Cox, *The Algebra of Probable Inference,* Johns Hopkins Press, Baltimore (1961).

31. J.M. Cozzolino, "The maximum-entropy distribution of the future market price of a stock," Op. Res. **6** (1973), 1200-1211.

32. R.G. Currie, "Evidence for 18.6 year M_n signal in temperature and drought conditions in North America since AD 1800," J. Geophys. Res. **86** (1981), 11055-11064.

33. G.J. Daniell and S.F. Gull, "Maximum entropy algorithm applied to image enhancement," Inst. Elec. Eng. Proc. **127E** (1980), 170-172.

34. R.N. McDonough, "Maximum-entropy spatial processing of array data," Geophys. **39** (1974), 843-851.

35. W.M. Elsasser, "On quantum measurements and the role of the uncertainty relations in statistical mechanics," Phys. Rev. **52** (1937), 987-999.

36. P.F. Fougere, "A solution to the problem of spontaneous line splitting in maximum entropy power spectrum analysis," J. Geophys. Res. **82** (1977), 1051-1054.

37. G.T. di Francia, "Capacity of an optical channel in the presence of noise," Opt. Acta 2 (1955), 5-8.

38. G.T. di Francia, "Resolving power and information," J. Opt. Soc. Am. **45** (1955), 497-501.

39. B.R. Frieden, "Statistical models for the image restoration problem," Comp. Graph. Image Process. 72 (1980), 40-59.

40. B.R. Frieden, *Probability, Statistical Optics, and Data Testing,* Springer, New York (1983).

41. W.F. Gabriel, "Spectral analysis and adaptive array superresolution techniques," Proc. IEEE **68** (1980), 654-666.

42. J.W. Gibbs, *Collected Works,* Yale University Press, New Haven (1928).

43. J.W. Gibbs, *Elementary Principles in Statistical Mechanics,* Dover, New York (1960).

44. W.T. Grandy, Jr., "Principle of maximum entropy and irreversible processes," Phys. Rep. **62** (1980), 175-266.

45. S. Guiasu, *Information Theory with Applications,* McGraw-Hill, New York (1977).

46. S.F. Gull and G.J. Daniell, "Image reconstruction from incomplete and noisy data," Nature **272** (1978), 686-690.

47. S. Haykin, B.W. Currie and S.B. Kesler, "Maximum-entropy spectral analysis of radar clutter," Proc. IEEE **70** (1982), 953-962.

48. S. Haykin, ed., *Nonlinear Methods of Spectral Analysis,* Springer, New York (1979).

49. D. Heath and W. Sudderth, "De Finetti's theorem on exchangeable variables," Am. Stat. **30** (1976), 188-189.

50. T.S. Huang, ed., *Picture Processing and Digital Filtering,* 2nd ed., Springer, New York (1979).

51. H. Hurwitz, Jr., "Entropy reduction in Bayesian analysis of measurements," Phys. Rev. A **12** (1975), 698-706.

52. R. Inguva and L.H. Schick, "Information theoretic processing of seismic data," Geophys. Res. Lett. **8** (1981), 1199-1202.

53. R. Inguva, "Stacking of noisy traces by maximum entropy methods," Conoco Report (1983).

54. D.D. Jackson, "The use of a priori data to resolve non-uniqueness in linear inversion," Geophys. J.R. astr. Soc. **57** (1979), 137-157.

55. E.T. Jaynes, "Information theory and statistical mechanics," Phys. Rev. **106** (1957), 620-630.

56. E.T. Jaynes, "Information theory and statistical mechanics. II," Phys. Rev. **108** (1957), 171-190.

57. E.T. Jaynes, "Information theory and statistical mechanics," in K. Ford, ed., *Statistical Physics,* Benjamin, New York (1963), 181-218.

58. E.T. Jaynes, "New engineering applications of information theory," in J.L. Bogdanoff and F. Kozin, eds., *Proceedings of the First Symposium on Engineering Applications of Random Function Theory and Probability,* Wiley, New York (1963), 163-203.

59. E.T. Jaynes, "Gibbs vs Boltzmann entropies," Am. J. Phys. **33** (1965), 391-398.

60. E.T. Jaynes, "Foundations of probability theory and statistical mechanics," in M. Bunge, ed., *Delaware Seminar in the Foundations of Physics,* Springer, Berlin (1967), 77-101.

61. E.T. Jaynes, "Prior probabilities," IEEE Trans. Syst. Sci. Cybern. **SSC-4** (1968), 227-241.

62. E.T. Jaynes, "The well-posed problem," Found. Phys. **3** (1973), 477-492.

63. E.T. Jaynes, "Where do we stand on maximum entropy?," in R.D. Levine and M. Tribus, eds., *The Maximum Entropy Formalism,* MIT Press, Cambridge, MA (1978), 15-118.

64. E.T. Jaynes, "The minimum entropy production principle," Ann. Rev. Phys. Chem. **31** (1980), 579-601.

65. E.T. Jaynes, "On the rationale of maximum-entropy methods," Proc. IEEE **70** (1982), 939-952.

66. E.T. Jaynes, *Papers on Probability, Statistics and Statistical Physics,* Reidel Publ. Co., Dordrecht (1983) – edited by R.D. Rosenkrantz.

67. E.T. Jaynes, "Prior information in inference," J. Am. Stat. Assn. (in press).

68. E.T. Jaynes, "Where do we go from here?," in C.R. Smith and W.T. Grandy, Jr., eds, *Maximum-Entropy and Bayesian Methods in Inverse Problems,* Reidel Publ. Co., Dordrecht (in press).

69. D.S. Jones, *Elementary Information Theory,* Oxford Univ. Press (1979).

70. D.H. Johnson, "The application of spectral estimation methods to bearing estimation problems," Proc. IEEE **70** (1982), 1018-1028.

71. R.W. Johnson and J.E. Shore, "Minimum cross-entropy spectral analysis of multiple signals," IEEE Trans. Acoust., Speech, Sig. Proc. **ASSP-31** (1983), 574-582.

72. S.M. Kay and S.L. Marple, Jr., "Spectrum analysis – a modern perspective," Proc. IEEE **69** (1981), 1380-1419.

73. R.J. Keeler, "Uncertainties in adaptive maximum entropy frequency estimators," IEEE Trans. Acoust., Speech, Sig. Proc. **ASSP-26** (1978), 469-471.

74. A.I. Khinchin, *Mathematical Foundations of Information Theory,* Dover, New York (1957).

75. R.T. Lacoss, "Data adaptive spectral analysis methods," Geophys. **36** (1971), 661-675.

76. J.S. Lim and N.A. Malik, "Maximum entropy spectrum estimation of signals with missing correlation points," IEEE Trans. Acoust., Speech, Sig. Proc. **ASSP-29** (1981), 1215-1217.

77. N.F.G. Martin and J.W. England, *Mathematical Theory of Entropy,* Addison-Wesley, Reading, Mass. (1981).

78. J.H. McClellan, "Multidimensional spectral estimation," Proc. IEEE **70** (1982), 1029-1039.

79. G. Minerbo, "MENT: A maximum entropy algorithm for reconstructing a source from projection data," Comp. Graph. Image Process. **10** (1979), 48-68.

80. T.C. Minter, Jr., "Minimum Bayes risk image correlation," SPIE **238** (1980), 200-208.

81. D. Montgomery, L. Turner and G. Vahala, "Most probable states in magnetohydrodynamics," J. Plasma Phys. **21** (1979), 239-251.

82. R.A. Monzingo and T.W. Miller, *Introduction to Adaptive Arrays,* Wiley, New York (1980).

83. W.I. Newman, "Extension to the maximum entropy method II," IEEE Trans. Inform. Theory **IT-25** (1979), 705-708.

84. J.P. Noonan, N.S. Tzannes and T. Costello, "On the inverse problem of entropy maximizations," IEEE Trans. Info. Th. **IT-22** (1976), 120-123.

85. D.W. North, "The invariance approach to the probabilistic encoding of information," Ph.D. Dissertation, Stanford University, Stanford, CA (1970).

86. H. Ogura and Y. Yoshida, "Spectral analysis and subtraction of noise in radar signals," IEEE Trans. Aerosp. El. Syst. **AES-17** (1981), 62-71.

87. D. Otero et al., "Ehrenfest theorem and information theory," Phys. Rev. A **26** (1982), 1209-1217.

88. A. Papoulis, "Maximum entropy and spectral estimation: a review," IEEE Trans. Acous., Speech, Sig. Proc. **ASSP-29** (1981), 1176-1186.

89. J.R. Pierce, *An Introduction to Information Theory,* 2nd ed., Dover, New York (1980).

90. E. Prugovecki, "Information-theoretical aspects of quantum measurement," Int. J. Th. Phys. **16** (1977), 321-331.

91. H.R. Radoski, P.F. Fougere and E.J. Zawalick, "A comparison of power spectral estimates and applications of the maximum entropy method," J. Geophys. Res. **80** (1975), 619-625.

92. J. Rissanen, "Modeling by shortest data description," Automatica **14** (1978), 465-471.

93. J. Rissanen, "A universal prior for integers and estimation by minimum description length," Ann. Stat. **11** (1983), 416-431.

94. E.A. Robinson and S. Treitel, "Maximum entropy and the relationships of the partial autocorrelation to the reflection coefficients of a layered system," IEEE Trans. Acoust., Speech, Sig. Proc. **ASSP-28** (1980), 224-235.

95. C.E. Shannon and W. Weaver, *The Mathematical Theory of Communication,* Univ. Illinois Press, Urbana (1962).

96. R. Shaw, ed., *Image Analysis and Evaluation,* Soc. Photo. Sci. Eng., Wash., D.C. (1977).

97. J.E. Shore and R.W. Johnson, "Axiomatic derivation of the principle of maximum entropy and the principle of minimum cross-entropy," IEEE Trans. Inform. Theory **IT-26** (1980), 26-37.

98. J.E. Shore, "Minimum cross-entropy spectral analysis," IEEE Trans. Acous., Speech, Sig. Proc. **ASSP-29** (1981), 230-237.

99. J.E. Shore, "Properties of cross-entropy minimization," IEEE Trans. Inform. Theory **IT-27** (1981), 472-482.

100. J.E. Shore, "Minimum cross-entropy pattern classification and cluster analysis," IEEE Trans. Pattern Anal. Mach. Intell. **PAMI-4** (1982), 11-17.

101. J.E. Shore, "Information theoretic approximations for M/G/1 and G/G/1 queuing systems," Acta Informatica **17** (1982), 43-61.

102. J. Skilling, A.W. Strong and K. Bennett, "Maximum-entropy image processing in gamma-ray astronomy," Mon Not. R. astr. Soc. **187** (1979), 145-152.

103. C.R. Smith, *The Maximum-Entropy Approach to Signal and Image Processing* (first draft of a monograph; based on lectures presented at the Univ. Alabama, Jan. 1983).

104. C.R. Smith and W.T. Grandy, Jr., eds., *Maximum-Entropy and Bayesian Methods in Inverse Problems,* Reidel Publ. Co., Dordrecht (in press).

105. D.E. Smylie, G.K.C. Clarke and T.J. Ulrych, "Analysis of irregularities in the earth's rotation," Meth. Comp. Phys. **13** (1973), 391-430.

106. C.P. Sonett, "Sunspot time series: spectrum from square law modulation of the Hale cycle," Geophys. Res. Lett. **9** (1982), 1313-1316.

107. L. Szilard, "On the decrease of entropy in a thermodynamic system by the intervention of intelligent beings," Behavioral Science **9** (1964), 301-310.

108. M. Tribus, *Thermostatics and Thermodynamics,* van Nostrand, New York (1961).

109. M. Tribus, *Rational Descriptions, Decisions, and Designs,* Pergamon, New York (1969).

110. H.J. Trussell, "The relationship between image restoration by the maximum a posteriori method and a maximum entropy method," IEEE Trans. Acous., Speech, Sig. Proc. **ASSP-28** (1980), 114-117.

111. H.J. Trussell, "Processing of x-ray images," Proc. IEEE **69** (1981), 615-627.

112. R.E. Turner and D.S. Betts, *Introductory Statistical Mechanics,* Sussex Univ. Press (1974).

113. N.S. Tzannes and J.P. Noonan, "The mutual information principle and applications," Inform. and Control **22** (1973), 1-12.

114. T.J. Ulrych, "Maximum entropy power spectrum of truncated sinusoids," J. Geophys. Res. **77** (1972), 1396-1400.

115. T.J. Ulrych et al., "Predictive filtering and smoothing of short records by using maximum entropy," J. Geophys. Res. **78** (1973), 4959-4964.

116. T.J. Ulrych and T.N. Bishop, "Maximum entropy spectral analysis and autoregressive decomposition," Rev. Geophys. Space Phys. **13** (1975), 183-200.

117. T.J. Ulrych and R.W. Clayton, "Time series modelling and maximum entropy," Phys. Earth and Planet. Int. **12** (1976), 188-200.

118. J.M. Van Campenhout and T.M. Cover, "Maximum entropy and conditional probability," IEEE Trans. Inform. Theory **IT-27** (1981), 483-489.

119. A. Wehrl, "General properties of entropy," Rev. Mod. Phys. **50** (1978), 221-260.

120. S.J. Wernecke and L. R. D'Addario, "Maximum entropy image reconstruction," IEEE Trans. Comp. **C-26** (1977), 351-364.

121. N. Wiener, *Cybernetics,* 2nd ed., MIT Press, Cambridge, Mass. (1961).

122. R. Willingale, "Use of the maximum entropy method in X-ray astronomy," Mon. Not. R. astr. Soc. **194** (1981), 359-364.

DEPARTMENT OF PHYSICS AND ASTRONOMY
UNIVERSITY OF WYOMING
LARAMIE, WYOMING 82071

RESEARCH DIRECTORATE
U.S. ARMY MISSILE LABORATORY, MICOM
REDSTONE ARSENAL, ALABAMA 35898

SIAM-AMS PROCEEDINGS
Volume **14**
1984

INVERSION AS LOGICAL INFERENCE — THEORY AND APPLICATIONS OF MAXIMUM ENTROPY AND MINIMUM CROSS-ENTROPY

John E. Shore

ABSTRACT. This paper takes the viewpoint that inversion problems are problems of inference. The paper reviews the theoretical foundations of maximum entropy and minimum cross-entropy inversion, and it demonstrates MCE inversion by solving some important spectrum analysis and image enhancement problems.

I. INTRODUCTION

The principles of maximum entropy (ME) and minimum cross-entropy (MCE) provide a general, information-theoretic method of inverting the equations

$$\int s_r(\mathbf{x}) q^\dagger(\mathbf{x})\, d\mathbf{x} = \bar{s}_r , \tag{1}$$

for known $s_r(\mathbf{x})$ and $\bar{s}_r$, $r=0, \cdots, M$. Both principles assume that $q^\dagger$ is a probability density function; MCE also takes into account a prior estimate of $q^\dagger$. In this paper, we review the theoretical foundations of ME and MCE inversion, and we demonstrate MCE inversion by solving some important image enhancement and spectrum analysis problems.

We take the point of view that inversion is an inference problem — from specific information about expected values of $q^\dagger(\mathbf{x})$, one wishes to infer the function $q^\dagger(\mathbf{x})$ itself. In seeking a general method of performing such inversions, one approach is to select or invent various possible methods and then study their properties. We take the opposite approach. That is, we define a minimum set of properties that we require of any reasonable inversion method, and we study the consequences for a particular class of methods. In particular, since we view inversion as an inference problem, we require that inversion methods satisfy requirements for consistent inference.

0080-5084/84 $1.00 + $.25 per page

Although (1) has a rather restricted form, many ME and MCE results also apply for the more-general, more-realistic constraints

$$\int s_r(\mathbf{x}) q^\dagger(\mathbf{x})\, d\mathbf{x} \le \bar{s}_r$$

and

$$\sum_{r=0}^{M} \alpha_r \left[\int s_r(\mathbf{x}) q^\dagger(\mathbf{x})\, d\mathbf{x} - \bar{s}_r \right]^2 \le \varepsilon^2 .$$

In the most general case, ME and MCE apply when the information about $q^\dagger$ restricts it to a convex set of probability densities. Although the assumption that $q^\dagger$ is a probability density may seem overly restrictive, often one can introduce an appropriate probability density into inversion problems whose original statements do not appear to be probabilistic. We do this in the examples considered in Section III.

II. MINIMUM CROSS-ENTROPY INVERSION

Let a system of interest take on states $\mathbf{x}$ according to the unknown probability density $q^\dagger(\mathbf{x})$. Suppose that $p(\mathbf{x})$ is a strictly positive *prior* estimate of $q^\dagger(\mathbf{x})$, and suppose that new information about $q^\dagger$ is obtained in the form of a set of known expected values (1). Let q be a *posterior* estimate of $q^\dagger$ — the result of inverting (1) in the context of the prior p. We require that q satisfy the *constraints* (1) and

$$\int d\mathbf{x}\, q(\mathbf{x}) = 1, \tag{2}$$

and we choose q by functional minimization,

$$H(q,p) = \min_{q'} H(q',p) , \tag{3}$$

where the minimization is carried out over the set of densities satisfying (1)-(2). For convenience, we refer to the general constraints (1)-(2) by the symbol I, and we introduce an operator $\circ$ that expresses (3) in the notation

$$q = p \circ I. \tag{4}$$

If $H(q,p)$ is a measure of the dissimilarity between q and p, then the generalized inversion (3) corresponds to selecting the density q that satisfies the constraints I and is otherwise as close as possible to the prior estimate p.

What should H be? One way to answer this question is to require that the operator $\circ$ satisfy certain axioms and then to derive consequences about H. Here is an informal description of one such set of axioms[1]:

I. *Uniqueness:* The result should be unique.

II. *Invariance:* The choice of coordinate system should not matter.

III. *System Independence:* It should not matter whether one accounts for independent information about independent systems separately in terms of different densities or together in terms of a joint density.

IV. *Subset Independence:* It should not matter whether one treats an independent subset of system states in terms of a separate conditional density or in terms of the full system density.

It is remarkable that the formal statement of these consistency axioms leads to a unique result[1]: $p \circ I$ must equal the result of minimizing the *cross-entropy* (discrimination information, directed divergence, Kullback-Leibler number, relative entropy),

$$H(q,p) = \int d\mathbf{x}\, q(\mathbf{x}) \log(q(\mathbf{x})/p(\mathbf{x})) \ . \tag{5}$$

The posterior $q = p \circ I$ has the form ([2, 3, 4])

$$q(\mathbf{x}) = p(\mathbf{x}) \exp\left[-\lambda - \sum_{r=0}^{M} \beta_r s_r(\mathbf{x})\right] , \tag{6}$$

where the β_r and λ are Lagrangian multipliers determined by (1) and (2). Conditions for the existence of solutions are discussed by Csiszàr [4].

Cross-entropy minimization can be viewed as a generalization of entropy maximization [5, 6]. — MCE reduces to ME when the prior is uniform [1]. MCE can also be viewed as equivalent to ME, with p accounting for a generalized transformation from a space in which the prior is uniform [7, 8]. The ME formalism is given by (1)-(6) with the prior deleted.

As a general method of statistical inference, MCE minimization was first introduced by Kullback [2], has been advocated in various forms by others [9, 7, 10, 11], and has been applied in various fields (for a list of references, see [1]). Recent successful applications include spectrum analysis [12, 13, 14, 15], pattern classification [16], speech coding [17], speech recognition [18, 19, 20], speech noise reduction [21], and queuing theory [22].

III. APPLICATION TO SPECTRUM ANALYSIS AND IMAGE ENHANCEMENT

In this section we apply MCE inversion to the equation

$$\sum_{k=1}^{N} s_{rk} O_k = D_r \qquad (r = 0, \cdots, M), \tag{7}$$

where D_r and s_{rk} are known, and where $M \leq N$. This inversion is basic to many important fields. Two in particular are the power spectrum estimation of stationary time series and the enhancement of two-dimensional images. In spectrum estimation,

D_r = autocorrelation value at lag t_r,

O_k = power at frequency f_k, and

s_{rk} = Fourier functions = $2\cos(2\pi f_k t_r)$.

In image enchancement,

D_r = image intensity at point r,

O_k = object intensity at point k, and

s_{rk} = point spread function.

If we write (7) in the matrix form $\mathbf{s} \cdot \boldsymbol{O} = \boldsymbol{D}$, it can in principle be solved by

$$\boldsymbol{O} = \mathbf{s}^{-1} \cdot \boldsymbol{D} \ . \tag{8}$$

In spectrum analysis applications, $M < N$ often holds; in this case, (7) has no unique solution. A large class of spectrum estimation methods proceed by extrapolating D_r so as to take on reasonable values in the region $M < r \leq N$, and then solving for the power spectrum by (8). In image enhancement, $M = N$ holds and a unique solution of (7) can be obtained in principle from (8). But D_r and s_{rk} usually are not known exactly, which causes (8) to be ill-behaved or misleading [23].

In this section we derive popular spectrum and image estimators by treating O_k as an expected value. For the spectrum estimator, O_k is the expected power at f_k (as in [24]); for the image estimator, O_k is the expected number of photons received at point k. We show that the difference between the estimators corresponds to different choices for the prior probability density.

A. Power Spectrum Estimation

We consider time-domain signals of the form

$$s(t) = \sum_{k=1}^{N} a_k \cos(2\pi f_k t) + b_k \sin(2\pi f_k t), \tag{9}$$

where the a_k and b_k are random variables and where the f_k are nonzero frequencies. Since any stationary random process $g(t)$ can be obtained as the limit of a sequence of processes with discrete spectra [25, p. 36], (9) is quite general. With suitable choices for the frequencies and amplitudes, the mean square error $E(|g(t)-s(t)|^2)$ can be made arbitrarily small.

We describe the random process $s(t)$ in terms of a probability density $q^\dagger(\mathbf{a},\mathbf{b})$, where $\mathbf{a} \equiv (a_1, \cdots, a_N)$ and $\mathbf{b} \equiv (b_1, \cdots, b_N)$. If we measure the power at frequency f_k, we should get

$$O_k = E(\tfrac{1}{2}(a_k^2+b_k^2)) = \int \tfrac{1}{2}(a_k^2+b_k^2) q^\dagger(\mathbf{a},\mathbf{b})\, d\mathbf{a}\, d\mathbf{b} . \tag{10}$$

As measurements, however, we have the D_r in (1) rather than the O_k. But substitution of (10) into (7) yields

$$\int \sum_{k=1}^{N} \tfrac{1}{2}(a_k^2+b_k^2) s_{rk}\, q^\dagger(\mathbf{a},\mathbf{b})\, d\mathbf{a}\, d\mathbf{b} = D_r , \tag{11}$$

which has the form of expected value constraints (1). We can therefore estimate $q^\dagger$ by $q = p \circ I$, where p is a suitable prior and I is the set of constraints (11). Posterior estimates for the O_k then follow from (10), with $q^\dagger$ replaced by q.

Given prior estimates P_k for the power at f_k, what is a suitable prior density $p(\mathbf{a},\mathbf{b})$? Clearly, it must satisfy

$$P_k = \int \tfrac{1}{2}(a_k^2+b_k^2) p(\mathbf{a},\mathbf{b})\, d\mathbf{a}\, d\mathbf{b} . \tag{12}$$

For time series (9), a reasonable choice for p is the multi-variate gaussian

$$p(\mathbf{a},\mathbf{b}) = \prod_{k=1}^{N} \frac{1}{2\pi P_k} \exp(-(a_k^2+b_k^2)/2P_k) . \tag{13}$$

The rest of the derivation is simpler in terms of the variable

$$x_k \equiv \tfrac{1}{2}(a_k^2+b_k^2). \tag{14}$$

Transforming (10), (11), and (13) from $(\mathbf{a},\mathbf{b})$ to $\mathbf{x}$ yields [26]

$$O_k = \int x_k q^{\dagger}(\mathbf{x})\, d\mathbf{x} , \tag{15}$$

$$\int \sum_{k=1}^{N} x_k s_{rk} q^{\dagger}(\mathbf{x})\, d\mathbf{x} = D_r , \tag{16}$$

and

$$p(\mathbf{x}) = \prod_{k=1}^{N} (1/P_k)\exp(-x_k/P_k) . \tag{17}$$

Given the prior (17) and the constraints (16) one can compute a minimum cross-entropy posterior estimate $q(\mathbf{x})$ of the form (6). The result is

$$q(\mathbf{x}) = e^{-\lambda}\prod_{k=1}^{N}\frac{1}{P_k}\exp\left[-\frac{x_k}{P_k} - \sum_{r=0}^{M}\beta_r s_{rk} x_k\right]$$

$$= \prod_{k=1}^{N}\left[\frac{1}{P_k}+\sum_r \beta_r s_{rk}\right]\exp\left[-\left(\frac{1}{P_k}+\sum_r \beta_r s_{rk}\right)x_k\right] , \tag{18}$$

where the β_r are Lagrangian multipliers determined by the constraints (16) [12]. The resulting minimum cross entropy spectrum analysis (MCESA) estimate of the power spectrum is just $O_k = \int x_k q(\mathbf{x})\, d\mathbf{x}$, namely

$$O_k = \frac{1}{\dfrac{1}{P_k} + \sum_{r=0}^{M}\beta_r s_{rk}} . \tag{19}$$

If one assumes a flat prior estimate of the power spectrum ($P_k = P$) and a zero-lag autocorrelation value ($t_0=0$), the prior in (19) is absorbed into the multiplier β_0 and the result reduces to that of Burg's maximum entropy spectral analysis (MESA) [27, 28, 12]. For equally spaced autocorrelation lags, $t_r = r\,\Delta t$, the MESA result takes on the familiar form

$$O_k = \frac{\sigma^2}{\left|\sum_{r=0}^{M} a_r z^{-r}\right|^2} , \tag{20}$$

where $z = \exp(-2\pi i f_k \Delta t)$, a_r are inverse filter sample coefficients, and σ^2 is a gain [12]. This is the well-known all-pole, autoregressive, or linear prediction form, which can also be derived by various equivalent formulations [29, 30, 31, 32]. It has become one of the most widely used spectrum analysis techniques in geophysical data processing

[33, 34, 35, 36] and speech processing [30, 17].

Equation (7) and the estimator (19) are unrealistic for at least two reasons. First, the data D_r usually include contributions from various independent sources (including background noise) that are not modeled explicitly. Second, (19) incorporates a prior estimate P_k, but the posterior estimate O_k does not depend on the extent to which P_k is considered to be a good estimate. If one believes only a little in P_k, one can always use a flat prior instead (MESA), but there should be some intermediate ground on which one can make use of the prior to a suitable extent. Generalizations of MCESA are available for both of these problems [13, 14]. In both cases, we assume that O is the sum of L independent signals and we generalize (7) to

$$\sum_{i=1}^{L}\sum_{k=1}^{N} s_{irk}\, O_{ik} = D_r \qquad (r = 0, \cdots, M)\,. \tag{21}$$

Here, $O_i \equiv O_{i1}, \cdots, O_{iN}$ is the power spectrum of the ith signal or noise source, $i = 1, \cdots, L$, and $s_{irk} = 2\cos(2\pi f_k t_r)$ for each signal, independent of i. Let P_i be a prior estimate of O_i. Furthermore, let w_{ik} be a frequency-dependent "weight" or "confidence" value that we assign to P_{ik}, in the sense that an increase in w_{ik} should result in the posterior O_{ik} being closer to P_{ik}, given that all other weights do not change. In these terms, the most general MCESA solution is

$$O_{ik} = \frac{1}{\dfrac{1}{P_{ik}} + \dfrac{1}{w_{ik}}\displaystyle\sum_{r=0}^{M}\beta_r s_{irk}}\,, \tag{22}$$

where the β_r are determined by the constraints (21) on the sum of the signals [13, 14].

B. Image Enhancement

We model images in terms of photon arrivals rather than in terms of the Fourier expansion (9). In particular, we write

$$O_k = \sum_{x_1=0}^{\infty} \cdots \sum_{x_N=0}^{\infty} x_k\, \mathbf{q}^{\dagger}(\mathbf{x}) \equiv \sum_{\mathbf{x}} x_k\, \mathbf{q}^{\dagger}(\mathbf{x})\,, \tag{23}$$

where x_k is the number of photons originating from object point k. Substitution of (23) into (7) yields

$$\sum_{\mathbf{x}}\sum_{k=1}^{N} x_k s_{rk}\, \mathbf{q}^{\dagger}(\mathbf{x}) = D_r\,, \tag{24}$$

which has the same form as (16) except that the integral in (16) is replaced by a sum. The difference between (16) and (24) is one of interpretation. In both spectrum and image estimation, we model O_k as an expected value. In the case of spectrum estimation, the variable x_k is the energy of a Fourier series term (see (9) and (14); in the case of image estimation, x_k is a photon count.

Given prior estimates P_k for the number of photons originating from the object point k, what is a suitable prior distribution $\mathbf{p}(\mathbf{x})$? Clearly it must satisfy

$$P_k = \sum_{\mathbf{x}} x_k \mathbf{p}(\mathbf{x}) , \tag{25}$$

which is analogous to (12). For images, a reasonable choice for p is the multivariate poisson distribution

$$\mathbf{p}(\mathbf{x}) = \prod_{k=1}^{N} e^{-P_k} \frac{(P_k)^{x_k}}{x_k!} . \tag{26}$$

Given the prior (26) and the constraints (24), one can compute a minimum cross-entropy posterior of the form (5). The result is

$$\mathbf{q}(\mathbf{x}) = e^{-\lambda} \prod_{k=1}^{N} e^{-P_k} \frac{(P_k)^{x_k}}{x_k!} \exp\left[- \sum_{r=0}^{M} \beta_r x_k s_{rk}\right]$$

$$= e^{-\lambda} \prod_{k=1}^{N} \frac{e^{-P_k}}{x_k!} \left[P_k \exp(- \textstyle\sum_r \beta_r s_{rk})\right]^{x_k} \tag{27}$$

where λ is chosen to satisfy the normalization constraint (2) and the β_r are chosen so that (26) satisfies the constraints (23). As a result of the normalization constraint, (26) becomes

$$\mathbf{q}(\mathbf{x}) = \prod_{k=1}^{N} e^{-O_k} \frac{(O_k)^{x_k}}{x_k!} , \tag{28}$$

where

$$O_k = \sum_{\mathbf{x}} x_k \mathbf{q}(\mathbf{x}) = P_k \exp\left[- \sum_{r=0}^{M} \beta_r s_{rk}\right] \tag{29}$$

is just the posterior estimate of object intensity. The β_r in (29) must be chosen so that the O_k satisfy (7). For a flat prior estimate $P_k = P$, (29) has the form

$$O_k = \sum_{\mathbf{x}} x_k \, \mathbf{q}(\mathbf{x}) = \exp\left[-\sum_{r=0}^{M} \beta'_r s_{rk}\right] , \tag{30}$$

where $\beta'_r = \beta_r - \log P$ are chosen to satisfy (7), since $\sum_r s_{rs} = 1$ holds for point spread functions. Equation (30) is equivalent to a popular "maximum entropy" estimator for image processing [37, 38, 39, 40].

Objections to (7) raised earlier in the case of spectrum estimation also apply to image enhancement. As was done for spectrum analysis, we replace (7) with (21). We interpret O_i is the intensity (photon count) distribution of the ith object and s_{irk} is its point spread function. The point spread function satisfies $\sum_r s_{irk} = 1$, and it may depend on i — for example, if the objects are located at different distances. The weights w_{ik} have the same interpretation as before. The resulting generalization of (29) is

$$O_{ik} = P_{ik} \exp\left[-\frac{1}{w_{ik}} \sum_{r=0}^{M} \beta_r s_{irk}\right] . \tag{31}$$

IV. SUMMARY

We considered ME and MCE as general methods of inversion, and we reviewed their theory from the viewpoint of consistent inference. As examples, we derived both image and spectrum estimators as applications of MCE. In both cases, the estimates O_k were treated as analogous expected values (see (15) and (23)). The different solutions corresponded to different choices for the prior — (17) and (36) have the same expected values but different functional forms. The appropriate prior for time series spectra lead to (19) and (22), while the appropriate prior for images lead to (29) and (31).

ACKNOWLEDGEMENTS

I thank R. Johnson for many helpful discussions and J. Buck for reviewing a draft of this paper.

REFERENCES

1. J. E. Shore and R. W. Johnson, "Axiomatic derivation of the principle of maximum entropy and the principle of minimum cross-entropy," *IEEE Trans. Inform. Theory* **IT-26**, pp.26-37 (Jan. 1980).

2. S. Kullback, *Information Theory and Statistics*, Dover, New York (1969). Wiley, New York, 1959

3. J. E. Shore and R. W. Johnson, "Properties of cross-entropy minimization," *IEEE Trans. Inform. Theory* **IT-27**, pp.472-482 (July 1981).

4. I. Csiszár, "I-divergence geometry of probability distributions and minimization problems," *Ann. Math. Stat.* **3**, pp.146-158 (1975).

5. E. T. Jaynes, "Information theory and statistical mechanics I," *Phys. Rev.* **106**, pp.620-630 (1957).

6. W. M. Elsasser, "On quantum measurements and the role of the uncertainty relations in statistical mechanics," *Phys. Rev.* **52**, pp.987-999 (Nov. 1937).

7. E. T. Jaynes, "Prior probabilities," *IEEE Trans. Systems Science and Cybernetics* **SSC-4**, pp.227-241 (1968).

8. J. E. Shore, "The choice of state spaces and prior estimates in cross-entropy minimization," in preparation.

9. I. J. Good, "Maximum entropy for hypothesis formulation, especially for multidimensional contingency tables," *Ann. Math. Stat.* **34**, pp.911-934 (1963).

10. A. Hobson and B. Cheng, "A comparison of the Shannon and Kullback information measures," *J. Stat. Phys.* **7**, pp.301-310 (1973).

11. R. W. Johnson, "Axiomatic characterization of the directed divergences and their linear combinations," *IEEE Trans. Inf. Theory* **IT-25**, pp.709-716 (Nov. 1979).

12. J. E. Shore, "Minimum cross-entropy spectral analysis," *IEEE Trans. Acoust., Speech, Signal Processing* **ASSP-29**, pp.230-237 (Apr. 1981).

13. R. W. Johnson and J. E. Shore, "Minimum-cross-entropy spectral analysis of multiple signals," *IEEE Trans. Acoust., Speech, Signal Processing* **ASSP-31**, to appear (June 1983).

14. R. W. Johnson and J. E. Shore, "Multi-signal minimum-cross-entropy spectrum analysis with weighted priors," NRL Report in publication, Naval Research Laboratory, Washington, D.C. (1983). (submitted to *IEEE Trans. Acoustics, Speech, Signal Processing.*)

15. R. W. Johnson, "Algorithms for single-signal and multisignal minimum-cross-entropy spectral analysis," NRL Report 8667, Naval Research Laboratory, Washington, D.C. (1983). (submitted to *IEEE Trans. Acoustics, Speech, Signal Processing.*)

16. J. E. Shore and R. M. Gray, "Minimum-cross-entropy pattern classification and cluster analysis," *IEEE Trans. Patt. Anal. and Machine Intell.* **PAMI-4**, pp.11-17 (Jan. 1982).

17. R. M. Gray, A. H. Gray, Jr., G. Rebolledo, and J. E. Shore, "Rate-distortion speech coding with a minimum discrimination information distortion measure," *IEEE Trans. Inform. Theory* **IT-27**, pp.708-721 (Nov. 1981).

18. J. E. Shore and D. Burton, "Discrete utterance speech recognition without time normalization — recent results," *Proceedings 1982 Int. Conf. Pattern Recognition*, pp.582-584, IEEE 82CH1801-0 (Oct. 1982).

19. J. E. Shore and D. Burton, "Discrete utterance speech recognition without time alignment," *IEEE Trans. Inform. Theory* (July, 1983).

20. J. E. Shore, D. Burton, and J. Buck, "A generalization of isolated word recognition using vector quantization," pp. 1021-1024 in *Proceedings ICASSP83*, Boston, MA (April, 1983). IEEE 83CH1841-6.

21. R. W. Johnson, J. E. Shore, D. Burton, and J. Buck, "Speech noise reduction by means of multi-signal minium cross-entropy spectral analysis," pp. 1129-1132 in *Proceedings ICASSP83*, Boston, MA (April, 1983). IEEE 83CH1841-6.

22. J. E. Shore, "Information theoretic approximations for *M/G/1* and *G/G/1* queuing systems," *Acta Informatica* **17**, pp.43-61 (1982).

23. B. R. Frieden, "Statistical models for the image restoration problem," *Computer Graphics and Image Processing* **12**, pp.40-59 (1980).

24. J. E. Shore and R. W. Johnson, "Which is the best entropy expression for speech processing — *-S log S* or *log S*?," NRL Report in publication, Naval Research Laboratory, Washington, D.C. (1983). (submitted to *IEEE Trans. Acoustics, Speech, Signal Processing*.)

25. Y. Yaglom, *An Introduction to the Theory of Stationary Random Functions*, Prentice-Hall, Englewood Cliffs, NJ (1962).

26. J. E. Shore, "Minimum Cross-Entropy Spectral Analysis," NRL Memorandum Report 3921, Naval Research Laboratory, Washington, D.C. 20375 (Jan. 1979).

27. J. P. Burg, "Maximum entropy spectral analysis," presented at the 37th Annual Meeting Soc. of Exploration Geophysicists, Oklahoma City, Okla. (1967).

28. J. P. Burg, "Maximum Entropy Spectral Analysis," Ph.D. Dissertation, Stanford University (1975). (University Microfilms No. 75-25,499)

29. A. VanDenBos, "Alternative interpretation of maximum entropy spectral analysis," *IEEE Trans. Inf. Theory* **IT-17**, pp.493-494 (July 1971).

30. J. D. Markel and A. H. Gray, Jr., *Linear Prediction of Speech*, Springer-Verlag, New York (1976).

31. S. M. Kay and S. L. Marple, Jr., "Spectrum analysis — a modern perspective," *Proc. IEEE* **69**, pp.1380-1419 (Nov., 1981).

32. A. Papoulis, "Maximum entropy and spectral estimation: a review," *IEEE Trans. Acoust., Speech, Signal Processing* **ASSP-29**, pp.1176-1186 (Dec., 1981).

33. R. T. Lacoss, "Data adaptive spectral analysis methods," *Geophysics* **36**, pp.661-675 (1971).

34. T. J. Ulrych and T. N. Bishop, "Maximum entropy spectral analysis and autoregressive decomposition," *Rev. Geophys. Space Phys.* **43**, pp.183-200 (1975).

35. D. E. Smylie, G. K. C. Clarke, and T. J. Ulrych, "Analysis of irregularities in the earth's rotation," *Methods in Computational Physics* **vol. 13**, pp.391-431, New York, Academic Press (1973).

36. E. A. Robinson, "A historical perspective of spectrum estimation," *Proceedings of the IEEE* **76**, pp.885-907 (Sept. 1982).

37. B. R. Frieden, "Restoring with maximum likelihood and maximum entropy," *J. Opt. Soc. Am.* **62**(4), pp.511-518 (April 1972).

38. R. Gordon and G. T. Herman, "Reconstruction of pictures from their projections," *Quarterly Bull. Center for Theor. Biol.* **4**, pp.71-151 (1971).

39. S. F. Gull and G. J. Daniell, "Image reconstruction from incomplete and noisy data," *Nature* **272**, pp 686-670 (April, 1978).

40. J. Skilling, "Maximum entropy and image processing -- algorithms and applications," *Proc. First Maximum Entropy Workshop* (1981).

Computer Science and Systems Branch
Information Technology Division
Code 7591
Naval Research Laboratory
Washington, D. C. 20375

SIAM-AMS PROCEEDINGS
Volume **14**
1984

PRIOR INFORMATION AND AMBIGUITY IN INVERSE PROBLEMS

E. T. Jaynes

ABSTRACT. Mathematically ill-posed problems, asking us to invert a singular or nearly singular operator, appear constantly in applications. Many attempts have been made to deal with such problems by inventing ad hoc algorithms which imitate the direct mathematical inversion that one would like to carry out.

But if we view these as problems of inference rather than inversion there is a formal Decision Theory that, by taking into account prior information and value judgments about the purpose of the inversion, can often guide us without ad hockery to an algorithm that is unique, numerically stable, and demonstrably optimal by some rather basic criteria of rational choice.

The method is illustrated by two simple examples, inversion of an integral equation in which the instability is resolved by prior information about the class of possible solutions; and inversion of a singular matrix (image reconstruction) in which the ambiguity is resolved by entropy factors; i.e., prior information about multiplicity.

1. ILL-POSED PROBLEMS. The terms "Well-Posed" and "Ill-Posed" are commonly attributed to Hadamard.[1] However, in the Nineteenth Century Bertrand[2] applied the epithet "mal posée" to his famous paradox in probability theory, in which one asks for the distribution of lengths of chords drawn at random on a circle. He evidently meant the term in the sense of "underdetermined".

In applications, underdetermined problems are the rule rather than the exception. In physics, engineering, or statistics it usually requires creative imagination (inventing models, or prior information such as initial conditions that we do not actually possess) to convert a real problem into one which is "well-posed" in the sense that the statement of the problem gives just enough information to determine one unique solution. In biology, econometrics, geophysical exploration, medical diagnosis, and synthesis of electrical filters or optical systems, a truly well-posed problem is virtually unknown.

We must therefore, of necessity, learn to reason somehow in logically indeterminate situations. G. Polya[3] termed this "plausible reasoning" and showed that even a pure mathematician uses it constantly. Polya's plausible reasoning remained qualitative, although he noted a loose correspondence with the relations of probability theory.

The quantitative use of probability theory for this purpose is what we shall call inference. As we use it, this term includes "Bayesian statistics", but is more general in that Bayes' theorem is only one of the useful principles of inference. Others presently known include group invariance, maximum entropy, and coding theory.

Examples of underdetermined problems are spectrum analysis, image reconstruction, determining the shape of a target from its radar reflections, determining crystallographic or macromolecular structure from X-ray scattering patterns; and as we shall see, the Statistical Mechanics of J. Willard Gibbs for predicting thermodynamic properties--in all cases from incomplete and/or approximate data.

A problem might be ill-posed, in the more general sense of "not well-posed", in other ways. Usually, an overdetermined problem would be considered not merely ill-posed, but wrongly posed, calling for reformulation rather than inference. But a problem may also be formally well-posed, but nevertheless pragmatically without a unique solution because of practical difficulties such as instability, that make it impossible to use the solution with real data.

Examples of such "morally ill-posed" problems are analytic continuation from numerical data, some Fredholm integral equations, extrapolating a solution of the diffusion equation backward in time, determining subsurface structure from surface gravimetric or seismic data, and the mechanics of billiard balls.

An unstable problem may be much like an underdetermined one, in that the formal solution must be supplemented by additional means, such as a preliminary smoothing or other "massaging" of the data in a particular way, or putting in a preliminary bias favoring some possibilities over others. The rationale for these is not always clear to the uninitiated, since it often arises out of prior knowledge of the subject-matter that is too extensive to be repeated in the statement of the problem, and can only be presumed "understood by the expert".

Furthermore, in both underdetermined and unstable problems we may require not only inference making use of expert prior information, but also value judgments indicating what we want the solution to accomplish, in order to arrive at useful results.

The above remarks sound very much like an introduction to Statistical Decision Theory, which shows us how to take prior information and value judgments into account, in a way that is proved optimal by some very fundamental and pretty nearly inescapable criteria of rational behavior. Indeed, the problem of inverting some singular or nearly singular operator A (i.e., given $y = Ax$, estimate x) would seem to call out for decision theory, just as clearly as the problem of driving a nail calls out for a hammer. This makes it curious that so little use has been made of this theory--or even probability theory--in dealing with ill-posed problems.

2. AD HOC ALGORITHMS. Tikhonov and Arsenin[4] explore a variety of unstable inverse problems, but do not see them as problems of inference or decision at all. Instead they invent various *ad hoc* algorithms that serve as a substitute for inversion; their "regularized" solutions force continuity in a neighborhood of a known exact solution, but in a way that does not necessarily make use of any prior information, or even any properties of the operator A--and for which it is therefore hard to give any convincing rationale.

When they take the nature of A into account to the extent of minimizing a mean square error metric, they have in effect rediscovered the Wiener[5] filter algorithm. But that too can become unstable in just the problems of greatest interest. In forecasting a time series with the Wiener prediction filter, for example, as we approach the limit where the Paley-Wiener criterion ceases to be satisfied, correlations persist for longer and longer times and formally the time series becomes more and more predictable. Actually, the prediction algorithm approaches analytic continuation and becomes less and less stable.

The limit is reached just for the physicist's favorite random function, Planck black-body radiation with power spectrum $I(\omega) \propto \omega^3/(e^{b\omega}-1)$. The Paley-Wiener integral $\int \log I(\omega)/(1+\omega^2)d\omega$ then diverges, and the conventional Wiener theory thus tells us that the time series (say, the x-component of electric field) is perfectly predictable from its past.

Of course, nobody familiar with the realities would believe this for an instant. The Wiener prediction algorithm here reduces to analytic continuation, not only impossible from numerical data, but even physically wrong for reasons apparent to physicists but neglected in the Wiener theory.

This reminds us that in many inverse problems, as the solution approaches instability, not only do our conclusions become highly sensitive to small changes in the data; they become equally sensitive to the exact physical assumptions underlying the theory itself. When the numerical algorithm becomes shaky, the whole foundation of the theory may also become shaky and a direct mathematical inversion, even if achieved, could be more misleading than useful.

Thus in a variety of real problems a different kind of philosophy and rationale is needed. Unique deductively obtained results (i.e., direct mathematical inversion) being impossible, we must set our sights on some other goal, perhaps more modest but attainable.

3. THE ROLE OF PROBABILITY THEORY. It appears to us that the reason for this neglect of inference/decision theory methods in dealing with ill-posed problems lies in the attitude toward probability theory itself that is instilled by most current pedagogy. As currently taught, probability theory does not seem

applicable unless a problem has some evident "element of randomness". Even then, the conventional "frequentist" interpretation places strong restrictions on the allowable forms of its application.

This was stressed by L. J. Savage[5] who noted that on the conventional view probabilities may be assigned only to "random variables" and not to hypotheses or parameters; and evidence as to the magnitude of a probability is to be obtained "by observation of some repetitions of the event, and from no other source whatsoever." But this means that (a) we are prohibited from using probability theory for inference about the very things about which we are most interested; and (b) we are prohibited from making use of any prior knowledge we might have--however cogent--if it does not happen to consist of frequency data.

As Savage and others have noted, such a program is almost never workable in problems of the real world and nobody can really adhere to it; yet it continues to be taught in most statistics courses.

However, we view probability theory, as did Good,[6] Savage[5], Jeffreys,[7] de Finetti[11] and many others, as basically a set of normative rules for conducting inference, with no necessary connection with the notion of "random variables". As we have expounded elsewhere,[9] the usual equations of probability theory are uniquely determined, as rules for inference, by some very elementary desiderata of consistency that make no reference to random experiments.

This broader view (actually, the original view of Jacob Bernoulli) is not in conflict with the conventional interpretation of probability as frequency in a random experiment; rather, the latter is included as a special case of probabilistic inference, for certain kinds of propositions or prior information (indeed, just the kind that the "frequentist" would want before using probability theory at all). But unlike the frequentist we consider it legitimate to assign probabilities to any clearly stated proposition; in the special case that it happens to be a proposition about frequencies, then the usual connections between probability and frequency are found to appear automatically, as a consequence of our theory. Bernoulli's "weak law of large numbers" was only the first of many such connections; another important one is contained in the famous de Finetti[8] exchangeability theorem.

We hope to show here that this reinterpretation of probability theory can convert an ill-posed problem of deductive reasoning into a well-posed problem of inference. Indeed, on the viewpoint advocated here, Bertrand's original "ill-posed" problem proves to be well-posed after all, with a unique solution that was conjectured by Borel and has been verified experimentally.[10]

In a sense, the following considerations might be called pre-mathematical rather than mathematical. At least in the applications we have studied, once a definition algorithm has been decided upon, realizing it explicitly tends to be straightforward. The difficulties that still hold up progress involve rather the preliminary rationale by which one decides which specific algorithm we should seek. That is the problem we address here.

4. A SIMPLE UNSTABLE PROBLEM. A proposal of Wolf and Mehta[12] to measure fluctuations in light intensity from data on fluctuations in the counting rate of the photoelectrons ejected by the light, gives us an excellent example of a problem of this type. It is formally well-posed and mathematically quite simple; but nevertheless exhibits almost all of the practical difficulties noted in the various more complicated inversion problems described at this Symposium. So let us see how to deal with it by inference rather than inversion.

The variable λ is the "light intensity" in units such that given λ, the conditional probability of observing n ejected photoelectrons in some nominal time interval, say a microsecond, is the Poisson distribution:

$$p(n|\lambda) = \exp(-\lambda)\lambda^n/n! \quad , \tag{1}$$

(More explicitly, $\lambda = qE/h\nu$, where ν = light frequency, E = light energy incident on the photocathode during that microsecond, and q = quantum efficiency).

But λ fluctuates from one microsecond to another according to some probability distribution $P(\lambda)$, and so the probability distribution for observed photoelectrons is a mixture of Poisson distributions:

$$p(n) = \int_0^\infty p(n|\lambda)P(\lambda)d\lambda \quad . \tag{2}$$

Wolf and Mehta note that this integral equation can be inverted, yielding the formal solution

$$P(\lambda) = \frac{1}{2\pi}\int_{-\infty}^{\infty} ds\ e^{-is\lambda} \sum_{n=0}^{\infty} (1+is)^n\ p(n) \tag{3}$$

and it seems at first glance natural to conclude, with them, that by use of (3) we can determine $P(\lambda)$ from experimental measurements of the distribution of observed counts n.

But anyone who tries to do this using for p(n) the observed counting distribution will discover three difficulties with the formal solution:

(A) With any finite amount of data there will be some $N = n_{max}$, the maximum number of photoelectrons observed in any microsecond. The sum in (3) is then a polynomial of degree N, and we obtain the startling conclusion that $P(\lambda)$ is a sum of derivatives of delta-functions:

$$P(\lambda) = \sum_{n=0}^{N} a_n \delta^{(n)}(\lambda) \quad . \tag{4}$$

Evidently, in order for this procedure to yield any acceptable distribution at all (i.e., a $P(\lambda)$ that is non-negative, normalized, and zero for $\lambda < 0$), the raw data must be extrapolated to $n \to \infty$ in a way that makes the sum in (3) a rather special, well-behaved analytic function of s. But the resulting $P(\lambda)$ will depend, not only on our data, but on how this extrapolation is carried out, and the statement of the problem does not seem to provide any criterion for preferring one extrapolation over another. This is an example where massaging of the data, guided by some of that expert prior information, is necessary.

(B) Even if we had an infinite amount of data, thus avoiding the extrapolation problem, we would not in general get an acceptable $P(\lambda)$, because the algorithm (3) has an unacceptable instability. A small change in the data, for example, $p_2 \to p_2 + 0.001$, $p_3 \to p_3 - 0.001$, would make an unbounded change in the right-hand side of (3). Common sense tells us that our conclusions ought to depend continuously on our data.

(C) Finally, we need to distinguish between the theoretically assigned probability p_n and the empirically measured frequency f_n. From the standpoint of frequentist probability theory, the observed f_n is equal to the "true" p_n plus some "random error" e_n. We would prefer to verbalize this a little differently, but the pragmatic result is the same: common sense tells us that, with incomplete data of finite accuracy, $P(\lambda)$ cannot be recovered with deductive certainty and perfect accuracy. We can make only rather crude estimates, and the accuracy of the estimate surely must depend on the amount of data we have. Yet the proposed solution (3) by inversion makes no reference at all to the amount of data or the accuracy of the result!

These same difficulties infect most of the inverse problems discussed at this Symposium and by Tikhonov and Arsenin. That these difficulties exist is, of course, well recognized; that their resolution by inference rather than inversion also exists, is our main message.

5. INFERENCE. It is clearly asking too much to expect that a finite amount of noisy data can be determine a full continuous function $P(\lambda)$ without further restrictions; and so we seek solutions only within some prescribed class of conceivable functions $P(\lambda|\theta)$, the different functions of the class being characterized by a parameter θ, which may be multidimensional. We then ask of the data only that they provide us with some "best" estimate of θ, and a statement about the reliability of the estimate. The particular class of functions considered would be chosen on the basis of some of that "expert prior information" about what kind of distributions are likely, given what one knows about the source of the light.

Fortunately, in the present problem little expertise is needed to see that, for most light sources the electric field $E(x,t)$ is a sum of an enormous number of small and nearly independent contributions (emission from individual atoms), and so by the Central Limit Theorem we expect a Gaussian distribution for E, therefore an exponential distribution for λ (which is essentially a space-time average of E^2) if the observation time is short compared to the coherence time.

An estimate of θ should, of course, be based on all the available evidence, and not just the evidence of one experiment or measurement. The same problem was faced by Laplace in the 18'th Century, as he sought to combine astronomical data from various sources into a single "best" estimate of some parameter, such as the mass of Saturn, appearing in the equations of celestial mechanics. He showed that probability theory can often tell us, uniquely, how evidence from different sources is to be combined into a final estimate, and what accuracy we are entitled to claim for that estimate.

In Laplace's problems, the final algorithm usually turned out to be weighted least squares. Unfortunately, then as now there was a tendency to confuse the algorithm with the method. In the nineteenth Century, losing sight of Laplace's rationale, "Least Squares" came to be considered a principle in its own right, to be applied indiscriminately in all problems whether or not it could be justified by the principles of probability theory. To avoid repeating past mistakes, it is important that in each new problem we re-examine the probabilistic basis for our algorithm.

In our present problem, the experiment consists of K repetitions of the measurement of n; denote the data obtained by $D \equiv \{n_1, n_2, \ldots, n_K\}$, and any additional prior information (which might be the result of previous experiments or a theoretical analysis) by I and let T stand for the statement: "θ is in the interval $(\theta, \theta+d\theta)$". The evidence I contains some information about θ, described by a probability $p(T|I)$. From the product rule of probability theory: $p(T,D|I) = p(T|D,I)p(D|I) = p(D|T,I)p(T|I)$ we have, if $p(D|I) > 0$ (i.e., the data set is a possible one),

$$p(T|D,I) = p(T|I)\,\frac{p(D|T,I)}{p(D|I)}\ . \tag{5}$$

Then, indicating a probability density $f(\theta|X)$ of θ conditional on any information X according to $p(T|X) = f(\theta|X)d\theta$, the final probability distribution for θ, given the prior information and the data, will have the form

$$f(\theta|D,I) = A\,f(\theta|I)\,p(D|\theta,I) \tag{6}$$

where A is a normalizing constant, independent of θ. The evidence contained in the experimental data D thus resides entirely in the θ dependence of the factor $p(D|\theta,X)$; all other details of the data are irrelevant for the estimation of θ. Usually, I will be relevant to the probability of obtaining the data D only through its relevance to the value of θ, in which case it is superfluous in this factor: $p(D|\theta,I) = p(D|\theta)$.

For example, suppose we seek the unknown distribution $P(\lambda)$ in the aforementioned class of exponential densities $P(\lambda|\theta) = \theta^{-1} \exp(-\lambda/\theta)$ corresponding to Gaussian distributions for the field components. θ is then the average light intensity; given θ, the probability of obtaining exactly n counts in any one measurement is

$$p(n|\theta) = \int_0^\infty p(n|\lambda)P(\lambda|\theta)d\lambda = \theta^n/(1+\theta)^{n+1} \tag{7}$$

and if successive measurements are statistically independent (i.e., separated by a time long compared to the correlation time of the light), the probability of obtaining the entire run of data D is a product of K such factors:

$$p(D|\theta) = \frac{\theta^N}{(1+\theta)^{N+K}} \tag{8}$$

where $N = \Sigma_i n_i$ is the total number of counts observed.

If the additional evidence I yields a probability density $f(\theta|I)$ which varies little in the range of θ where (8) is appreciably large, then I is very uninformative compared to the evidence of the experiment (i.e., the experiment is well-designed); and from (6) the final probability density $f(\theta|D,I)$ will be, for all practical purposes, proportional to (8) in its dependence on θ.

However, a slight formal refinement may be achieved by considering the prior probability density $f(\theta|I)$ a little more carefully; and if we have very little data the difference might be noticeable. Suppose we wish to express "complete prior ignorance" of the value of θ; what function $f(\theta|I)$ does this? Stated in this way, the question has been rightly rejected in the past as ill-posed; the phrase "complete ignorance" is too vague to define any specific mathematical problem. But in fact we are not completely ignorant; if we know the distribution $P(\lambda|\theta)$ we can hardly be ignorant of the fact that θ is a scale parameter.

Presumably, by ignorance of the absolute scale of the problem one ought to mean a state of knowledge that is not changed by a small change in that absolute scale; just as ignorance of one's location is a state of knowledge that is not changed by a small change in that location. One may, therefore, view "ignorance" as an invariance property; the probability density that is invariant under the group of scale changes ($\theta \to \theta' = a\theta$) satisfies the functional equation $f(\theta) = af(a\theta)$; i.e., it is $f(\theta|I) = (1/\theta)$.

Indeed, this prior was advocated long ago, on partly intuitive grounds, by Jeffreys.[7] The group invariance argument[13] is at least a strong heuristic principle taking a step toward a more rigorous derivation, but it still depends on intuition to the extent that the user must choose the group. We have now taken another step in proving, via the integral equations of marginalization theory,[14] that in a problem with two parameters (θ,α) with θ a scale parameter, the Jeffreys prior $(1/\theta)$ is uniquely determined as the only prior that is "completely uninformative about α" without further qualifications.

Because of this convergence of quite different lines of argument, and the good pragmatic success we have had in using it for many years, we shall advocate using the Jeffreys prior here if we wish to express a completely open prior opinion, that leaves the entire decision to the evidence of the data (thus achieving R. A. Fisher's goal of "letting the data speak for themselves").

The fact that the Jeffreys prior is improper (i.e., not normalizable) could be dealt with, if needed, by approaching it as the limit of a sequence of proper priors, as we have shown elsewhere.[14] However, in the present problem the integrals converge so well that this is not necessary; our result is the same.

If we do have cogent prior information and relatively little data, then adopting a different prior distribution which expresses that prior information may improve the reliability of our estimates. This has been found particularly in recent work on forecasting economic time series,[15] where incorporating prior information about regression coefficients can make a quite noticeable improvement in the forecasts; and in the problem of seasonal adjustment,[16] where prior information about the smoothness of the seasonal component can make a major change in our estimate of the irregular component.

With the Jeffreys prior, (6) and (8) yield the posterior density

$$f(\theta|D,I) = \frac{\Gamma(N+K)}{\Gamma(N)\Gamma(K)} \frac{\theta^{N-1}}{(1+\theta)^{N+K}} , \quad 0 < \theta < \infty \tag{9}$$

and it is usually sufficient to express our conclusions in the form of a few moments or percentiles of this distribution. To find the percentiles, note that (9) is a Beta distribution in the variable $x \equiv \theta/(1+\theta)$, so that the identity of the incomplete Beta function and the incomplete Binomial sum[17] gives the cumulative distribution

$$p(\theta<\alpha|D,I) = \sum_{r=0}^{K-1} \frac{\Gamma(N+r)}{\Gamma(N)\, r!} \frac{\alpha^N}{(1+\alpha)^{N+r}} \tag{10}$$

in a form for computer evaluation (the "Snedecor F-tables" of the statistician could be used also, for the particular percentiles tabulated).

The moments of the distribution (9) are found to be

$$E(\theta^m|D,I) = \frac{\Gamma(N+m)\Gamma(K-m)}{\Gamma(N)\Gamma(K)} , \quad -N < m < K \quad . \tag{11}$$

The estimate of θ which minimizes the expected square of the error (our "value judgment" for this case) is then $t \equiv E(\theta|D,I) = N/(K-1)$, and the variance of (9) is

$$\sigma^2 = t(t+1)/(K-2) \tag{12}$$

The (mean) $\pm$ (standard deviation)

$$(\theta)_{est} = t \pm \sigma \tag{13}$$

is then a reasonable statement of our "best" estimate and its accuracy. For example, if we wish to determine θ to $\pm 1\%$ accuracy for a light source that gives a counting rate of about t counts/microsecond, we shall require a number of observations $K > 10^4 (1 + t^{-1})$, a result which is hardly surprising to common sense, but of which the attempted inversion (3) gives no hint. As $(N,K) \to \infty$, (9) goes into a normal distribution: $\theta \sim N(t,\sigma)$.

This example shows how the inference (9) can answer, successfully, a more modest question than the direct inversion (3) tried to answer, unsuccessfully. Our conclusions are numerically stable, and the way in which the accuracy of those conclusions depends on the amount of data is now exhibited explicitly in (12).

Note, however, that in answering a more modest question, inference is not giving us any less information than inversion; for when a reliable inversion is possible, the likelihood factor $p(D|\theta,I)$ in (6) develops a single sharp peak and inference will reduce to inversion. An unstable inversion, with the superficial appearance of giving more information, is actually giving false/unreliable information without warning us of that fact. Inference gives us those conclusions that are actually justified by the prior information and data, and it tells us, by the probable error σ, how reliable our estimates are. It may also take into account prior information that inversion ignores.

We note how the appearance of an unstable geophysical inverse problem might be changed by a similar approach. Suppose we wish to infer some sub-surface property Q(z) (density, conductivity, elastic constants) from surface data D (gravimetric, electromagnetic, seismic). The data depend on Q(z) through some relation expressing physical law (potential theory, electromagnetic or acoustical wave equations, etc.); abstractly,

$$D = AQ(z) + N \tag{14}$$

where A is some operator, presumed known and N is whatever "noise"--unavoidable, uncontrollable, and unknown--places the ultimate limit on the accuracy of D.

The trouble is that the effect of Q(z) on D falls off, often exponentially fast, with z. Any attempt to reconstruct Q(z) by direct inversion of (14) must then become increasingly unstable and unreliable with depth. An inversion algorithm may be at least usable--although with unknown accuracy--up to a certain depth, beyond which it fails entirely.

If the problem were treated by inference the final result would be, not a single value for each depth z, but a family of probability densities parameterized by z:

$$F_z(Q) \equiv f(Q|z,D,I) \propto f(Q|z,I)p(D|Q,z,I) \tag{15}$$

such that $F_z(Q)\,dQ$ is the probability that, at depth z, Q lies in (Q,Q+dQ). As we go to greater depths and the estimates become, necessarily, less accurate, $F_z(Q)$ would indicate this by becoming broader. At depths where the data can give no information, $F_z(Q)$ would reduce to the prior density $f(Q|z,I)$. At depth z, our estimate will have the form

$$(Q)_{est} = t(z) \pm \sigma(z) \tag{16}$$

in which t(z) is our "best" estimate, that inversion had tried to give, and $\sigma(z)$ indicates how reliable that estimate is. As z increases, $\sigma(z)$ would increase from values small enough to make the estimate t(z) useful, to values so large that the data have told us nothing beyond whatever prior information we had. But the algorithm is stable, the increase is smooth and continuous, and there is no point at which the method suddenly fails.

It appears to us that conclusions stated in the format (15), (16) would indicate, more usefully and more honestly, what the data actually have to tell us about the question being asked.

6. GENERALIZED INVERSE PROBLEMS. In another class of problems arising constantly in applications, the trouble is not merely that the inversion is unstable; it is in principle impossible because the operator A is singular. Here too there have been unceasing efforts to resolve the ambiguity by inventing _ad hoc_ algorithms that imitate inversion, but take no note of the principles of inference.

For example, given values of a function

$$y(t) = \int d\omega Y(\omega) e^{i\omega t} \tag{17}$$

over only a part of its support, estimate its fourier transform $Y(\omega)$. The most obvious algorithm (take the transform of the data) gives us the fourier transform $Y_T(\omega)$ of a function $y_T(t)$ that is truncated to zero outside the measured region. In almost all real cases this would be arbitrary and unrealistic, and in some cases it would be unacceptable because it contradicts our prior information. Thus if $y(t)$ is the autocorrelation function of a time series, $Y(\omega)$ is its power spectrum, by definition non-negative. But as Burg[17] has emphasized, $Y_T(\omega)$ is not in general non-negative.

It is clear that there is a fundamental ambiguity here, since the data cannot distinguish between two estimates $Y_1(\omega)$, $Y_2(\omega)$ whose fourier transforms $y_1(t)$, $y_2(t)$ differ only outside the measured region. For a choice between them, one must appeal to prior information and/or value judgments.

To formulate problems of this type in a general, abstract way, there is an unknown "state of Nature" x, which for brevity and with a view to image reconstruction, we shall call "the scene". It might be a number, a vector, or a function. Intuitively (and even this step may require some of that creative imagination) we think of x as belonging to some set $X = \{x_1 \ldots x_n\}$ of possible scenes.

We would like to know the true scene x, but our information is incomplete. Instead, we know only the "blurred scene"

$$y = Ax \tag{18}$$

where A is an operator, supposed known but noninvertible. That is, we cannot recover x in the manner $x = A^{-1}y$ because the data y cannot distinguish between two scenes x, x' that satisfy the "homogeneous equation" $Ax - Ax' = 0$ (a true homogeneous equation if A is linear). The best we can do is to make an inference, in which we choose some estimate of x from our data:

$$\hat{x} = Ry \tag{19}$$

where R is a "resolvent" operator to be chosen. The conceptually difficult pre-mathematical problem is: by what criterion do we choose R?

It appears that deductive logic is able to give us only one restriction on R. Given the data (18), we know at least that x must lie in the class C (subset of X) of scenes x_i that satisfy $y = Ax_i$. Thus for all possible x we should have $y = Ax = A\hat{x} = ARy = ARAx$, or

$$ARA = A \tag{20}$$

and so R, in order not to conflict with deductive reasoning, must be a generalized inverse operator. Stated differently, as seen through the "distorting window" A, the estimated scene $\hat{x}$ should be indistinguishable from the true scene x.

A problem with this simple logical structure, in which A is considered known exactly, and the data are noiseless, will be called a <u>pure generalized inverse</u> problem. To have the data contaminated with noise or A unknown makes the problem "impure" in our terminology.

Even in the seemingly straightforward pure problem, there have been conceptual difficulties so serious that the logically necessary condition (20) is not always recognized; the literature of spectrum analysis, image reconstruction, and quality control contains various proposed algorithms that violate it.

Then what prior information and value judgments are available to guide us to one specific choice within C? No general answer can be given once and for all; basically, this must be pondered separately for each new problem, and the following suggestions surely will not be applicable in all cases. Yet there is a large class of problems in which a single "new" principle and a rather well-developed formalism resolve this ambiguity, in a way that proves to have demonstrable optimality properties and pragmatic success. With better understanding of its rationale[18] this class is growing, and its ultimate limits are not yet in sight.

In many problems, it develops that we actually have some highly cogent prior information of which most of us are hardly consciously aware, because--as our quotation from L. J. Savage shows--conventional probability theory adjures us to ignore it. But this adjuration is just the reason why "orthodox" statistics is incapable of dealing with pure generalized inverse problems.

Although, as we have shown elsewhere,[9,18] the following rationale applies without essential change to many different kinds of problems, it will suffice here to consider a finite, discrete, linear version which amounts to inverting a singular matrix. Our general "scene" x is then represented by a set of "true but unknown" numbers $\{x_1 \ldots x_n\}$ which we wish to estimate, the general "data" y by a smaller set $\{y_1 \ldots y_m\}$ of observations, $m < n$, the general operator A by a known $(m \times n)$ matrix:

$$y_j = \sum_{i=1}^{n} A_{ji} x_i \quad , \quad 1 \le j \le m \quad . \tag{21}$$

The resolvent operator R might, conceivably, be an $(n \times m)$ matrix; but this is not required. Indeed, if we restrict R to be linear we shall hardly get past the Wiener filter type of algorithm. We advocate below a highly nonlinear R, whose performance could not be matched by any linear operation.

At present, then, there are an infinite number of different operators R, linear and nonlinear, which all satisfy the necessary condition (20), and therefore yield estimates $\{\hat{x}_1 \ldots \hat{x}_n\}$ in the class C of possible scenes.

If we have no prior information about the phenomenon being observed, which would make some scenes in C inherently more likely than others, then it appears to us that the ambiguity is fundamentally irremediable and there can be no justification for any algorithm that picks out only one scene. In that case, the only honest "solution" to the problem would seem to be:

specify the entire class C, which contains (n-m) arbitrary parameters.

Suppose, however, that we know (or wish to adopt as a working hypothesis) that Nature is generating the scene x by N repetitions of some process (call it a "random experiment" if you like) which can, at each trial, produce any one of n results $\{r_1 \dots r_n\}$. In image reconstruction we might think of the scene as produced by distributing N little "elements of luminance" over the n pixels of the scene, such that the i'th pixel receives a total of N_i elements, and taking x_i as the fraction $x_i = N_i/N$ of total luminance in the i'th pixel. It is much like tossing N pennies onto a floor whose square tiles are numbered 1 to n, and noting how many land on the ith tile.

But this picture of the mechanism constitutes relevant prior information; there are _a priori_ n^N different conceivable things that could happen in this sequence of tosses, and of these a given scene x could be realized in a number of ways given by the multiplicity factor

$$W(\text{scene}) = \frac{N!}{(Nx_1)! \dots (Nx_n)!} \quad . \tag{22}$$

For large N the Stirling approximation gives asymptotically

$$\frac{1}{N} \log W(\text{scene}) \sim -\sum_i x_i \log x_i = H(\text{scene}) \quad , \tag{23}$$

the Shannon entropy of that scene. For all practical purposes, then, we may take the multiplicity of a scene as

$$W(\text{scene}) = e^{NH(\text{scene})} \quad . \tag{24}$$

With this observation, the ambiguity of our inversion problem is resolved. Scenes of higher entropy are inherently more likely because they have higher multiplicity; i.e., they can be realized by Nature in more ways. The scene which has maximum entropy subject to the constraints (21) is the one with the greatest multiplicity of all those in the class C of scenes permitted by our data; and so unless we have further prior information not yet brought to bear on the problem, it would seem irrational to choose any estimate other than the scene of maximum entropy.

There remain the questions of uniqueness and sharpness of this result. For most purposes uniqueness is disposed of by noting that the set $\{S_h : H > h\}$ of scenes with entropy greater than h is, by well-known properties of entropy, strictly convex. A sufficient (stronger than necessary) condition for uniqueness of the maximum-entropy point is then that the set C picked out by our constraints be convex, as is evidently the case for the constraints (21).

Our solution point is then a point of tangency of the set C with one of the sets S_h.

The sharpness of the result is indicated by the Entropy Concentration Theorem[18] presented recently in some detail; suffice it to say here that if we count the scenes by their multiplicities, then not only is the scene of maximum entropy favored over all others, for large N the overwhelming majority of all possible scenes have entropy very close to the maximum. For example, asymptotically, 99% of all scenes in class C have entropy in the range

$$H_{max} - \Delta H \leq H(scene) \leq H_{max} \quad (25)$$

where $H = \chi_k^2(0.01)/2N$, and $\chi_k^2(q)$ is the critical Chi-squared statistic at the 100q% significance level, for $k = n - m - 1$ degrees of freedom.

The analytical solution of this maximization problem is well known; define the partition function

$$Z(\lambda_1 \dots \lambda_m) \equiv \sum_i \exp(-\lambda_1 A_{1i} \dots -\lambda_m A_{mi}) \quad . \quad (26)$$

Then the maximum-entropy scene is given by

$$\hat{x}_i = Z^{-1} \exp(-\lambda_1 A_{1i} - \dots - \lambda_m A_{mi}) \quad (27)$$

in which the λ's (Lagrange multipliers in the constrained maximization) are chosen to fit the data (21).

To give further details here would duplicate what is in the presentations of J. Shore and J. Skilling at this Symposium. Another recent application to crystallographic inversion is given by Wilkins, et al.[19] We close with the observation that (26) is nothing but a generalized Gibbsian canonical distribution, which has been the basis of Statistical Mechanics for some 60 years. From the inference point of view, therefore, Statistical Mechanics was, historically, the first example of a pure generalized inverse problem in which the ambiguity was resolved by entropy maximization. There was no necessary connection with thermodynamics; but unfortunately, the generality of Gibbs' method was concealed by attempts to put frequentist interpretations on it, and it is only in very recent years that we have realized how much we still had to learn from Gibbs. The overwhelming "preference of Nature" for scenes of high entropy indicated by the Entropy Concentration Theorem is just what we have been calling the "Second Law of Thermodynamics" for a Century. The maximum-entropy methods expounded at this Symposium are only new applications of the Second Law, generalized beyond its original domain.

BIBLIOGRAPHY

1. J. Hadamard, Lectures on Cauchy's Problem, Yale University Press, New Haven, 1923.

2. J. Bertrand, Calcul des probabilités, Gauthier-Villars, Paris, 1889.

3. G. Polya, Mathematics and Plausible Reasoning, 2 Vols., Princeton University Press, Princeton, 1954.

4. A. N. Tikhonov & V. Y. Arsenin, Solutions of Ill-Posed Problems, V. H. Winston & Sons, Washington, D.C., 1977

5. L. J. Savage, The Foundations of Statistics, J. Wiley & Sons, Inc., New York, 1954.

6. I. J. Good, Probability and the Weighing of Evidence, C. Griffin & Co. Ltd., London, 1950.

7. H. Jeffreys, Theory of Probability, Oxford University Press, 1939.

8. B. de Finetti, "Prevision: its Logical Laws, its Subjective Sources", Translated from the French in H. E. Kyburg & H. I. Smokler, Studies in Subjective Probability, 2nd Edition, J. Wiley & Sons, Inc., New York, 1981.

9. E. T. Jaynes, "Where do we Stand on Maximum Entropy?", in R. D. Levine & M. Tribus, Eds., The Maximum Entropy Formalism, MIT Press, Cambridge, MA, 1978.

10. E. T. Jaynes, "The Well-Posed Problem", Foundations of Physics 3, (1973), 477-492.

11. B. de Finetti, The Theory of Probability, 2 vols., J. Wiley & Sons, Inc., New York, 1974.

12. E. Wolf & C. L. Mehta, "Determination of Statistical Properties of Light from Photoelectric Measurements", Phys. Rev. Lett. 13 (1964), 705-707.

13. E. T. Jaynes, "Prior Probabilities", IEEE Trans. Syst. Sci. Cybern, SSC-4 (1968), 227-241.

14. E. T. Jaynes, "Marginalization and Prior Probabilities", in A. Zellner, Ed., Bayesian Analysis in Econometrics and Statistics, North-Holland Publishing Co., Amsterdam, 1980, pp. 43-78.

15. R. B. Litterman, "A Bayesian Procedure for Forecasting with Vector Autoregression", Ph.D. Thesis, University of Minnesota, 1979.

16. E. T. Jaynes, "Highly Informative Priors", Proceedings of the Second International Meeting on Bayesian Statistics, Valencia, University of Valencia Press (in press).

17. J. P. Burg, "Maximum Entropy Spectrum Analysis", Ph.D. Thesis, Stanford University, 1975.

18. E. T. Jaynes, "On the Rationale of Maximum Entropy Methods", Proc. IEEE, 70 (1982), 939-982.

19. S. W. Wilkins, J. N. Varghese, and M. S. Lehmann, "Statistical Geometry: A Self-Consistent Approach to the Crystallographic Inversion Problem Based on Information Theory", Acta Cryst. A39 (1983), 47-60.

DEPARTMENT OF PHYSICS
WASHINGTON UNIVERSITY
ST. LOUIS, MISSOURI 63130

SIAM–AMS PROCEEDINGS
Volume 14
1984

THE ENTROPY OF AN IMAGE

J. Skilling S.F. Gull

ABSTRACT. We investigate how entropy should be used in image reconstruction and in the analysis of time-series and particle distributions. We find that one must always use the Shannon/Jaynes formula $-\sum p \log p$ when attempting to reconstruct the shape of an image or the power spectrum of a time-series. The alternative Burg formula $\sum \log W$ was derived and is correct for predicting samples from a time-series, but we give theoretical and practical grounds for rejecting its use in image reconstruction and power spectrum analysis. Finally we generalise the simple Shannon/Jaynes approach to encompass new and powerful techniques of data analysis.

1. INTRODUCTION

Maximum entropy is being increasingly used as a technique of image reconstruction using data of many different types. The straightforward linear problems of reconstruction from Fourier transform data[1], optical or X-ray convolution data[1-11], tomographic data[12,13] and the like are now being supplemented by nonlinear applications such as phaseless Fourier data[14,15] and "blind" deconvolution. The extensive demonstrable success of maximum entropy over and above conventional techniques[11,16] demands a clear understanding of the fundamental theory underlying the technique.

In general terms, the problem of image reconstruction is fundamentally ill-posed. Only a selection of data from an external object is observed, and even the data one does have are corrupted by noise. It follows that there can be very many "feasible images" which are consistent with one's data. This "feasible set" contains a complete degree of freedom for each independent unmeasured parameter, and allows a range of values for each measured but noisy parameter. From an austere falsificationist point of view this feasible set is the result of the experiment: images lying outside the set can be excluded. For sufficiently small problems, such as are encountered in elementary statistics with the estimation of only one or maybe two parameters x, one can use the feasible

---

1980 Mathematics Subject Classification. 62-07, 62M99, 62M15

set directly and display results in the familiar form of a confidence interval $a \leq x \leq b$.

With an image this is simply not possible. Ideally, an image is spatially continuous, but even when divided into finite pixels it may have a million separate intensities to be determined, and the austere approach become wholly impractical. It is a matter of practical necessity to select some small number of images (ideally just one) from the feasible set and present this as the "result of the experiment", with the open acknowledgement that one is merely presenting a deliberate selection.

The maximum entropy method consists of choosing that single feasible image which has greatest entropy

$$S = - \sum_i p_i \log(p_i/m_i)$$

where p_i is a probability corresponding to an appropriate quality of the image, and m_i is the corresponding measure. We refer to this as the Shannon/Jaynes entropy.

2. ENTROPY

The concept of entropy is intimately connected with probability distributions, and traditionally entropy has been derived from within probability theory[17-20]. Entropy measures one's degree of uncertainty about the answers to a question. It has a technical interpretation as the measure of uncertainty in a probability distribution function (pdf). Specifically, the entropy is the number of bits of information one expects to require in order to localise a random sample from the pdf. However, the pdf need not be based on physical frequencies. Thus an external object may be unique in its own right, and not be a sample from any physical ensemble, even though our partial knowledge of it may be encoded as a pdf.

Given imperfect data which constrain but do not fully determine a pdf, Jaynes[20] has suggested that one would be wise to use the least committal (maximum entropy) pdf in predictions of future samples. Taking these ideas further, Shore[21] has shown that the maximum entropy selection is the only choice consistent with simple invariance conditions.

It is important to remember that even a single object may have many different probability distributions associated with it, each having its own entropy. For example, this paper has a typographical entropy of its symbols, a locational entropy in the literature, a truth entropy of its contents, and so on.

This simple fact has caused much confusion. The concept of entropy has been applied, inter alia, to fields as diverse as time-series, particle distributions and image reconstruction. Different final formulae arise in each case, which is perfectly natural since distinctly different questions are being asked in each field. It is most certainly not the case that formulae in one field can be naively transferred to another. We proceed to investigate this in more detail.

3. TIME-SERIES: THE PREDICTION PROBLEM

Suppose, when gambling \$1, one's payoff x has been observed to have a long-term mean $\mu = \langle x \rangle$ of 80¢. What is the pdf p(x) of payoff x? There are many feasible p(x) having the correct mean $\sum xp(x) = 80$, each having its own entropy. (To complete the definition of entropy, we assign unit measure to each 1¢ digitisation of x, ignoring for simplicity any coarser digitisation into 10¢ or 25¢ multiples.)

A particular feasible p(x) is $\{p=1 \text{ if } x=80, \; p=0 \text{ otherwise}\}$. This certainly has the correct mean, but is highly committal about the individual results x. Its entropy, in fact, is zero. Another feasible p(x) is $\{p=0.5 \text{ if } x=0 \text{ or } x=160, \; p=0 \text{ otherwise}\}$. This is a bit less committal; its entropy is $S = \log 2$ which represents one bit of information. Less committal again, though it still prohibits any payoff above \$1.60, is $\{p=1/(2\mu+1) \text{ if } 0 \leq x \leq 2\mu, \; p=0 \text{ otherwise}\}$, for which the entropy is $S = \log(2\mu+1) = 7.3$ bits.

The pdf which is least committal about future samples is that which needs most bits of information to localise a particular sample x from the pdf p(x). It is the pdf which imposes least prejudice about the outcomes, and is by definition the maximum entropy pdf. For the gambling example, maximising

$$S = -\sum p(x) \log p(x)$$

under $\sum p(x) = 1$ and $\sum xp(x) = \mu = 80$ straightforwardly leads to the Bose distribution

$$p(x) = \left(\frac{1}{\mu+1}\right)\left(\frac{\mu}{\mu+1}\right)^x$$

with entropy

$$S = (\mu+1)\log(\mu+1) - \mu \log\mu = 7.8 \text{ bits.}$$

In the continuum limit $\mu \to \infty$, these formulae become

$$p(x) = \mu^{-1} e^{-\mu x}, \qquad S = \log(e\mu) .$$

Jaynes' suggestion[20] that one would be wise to use the maximum entropy pdf in prediction of future samples is seen to be eminently sensible.

More complicated examples can also be analysed. Thus one could observe the mean $\mu = \int xp(x)dx$ and variance $\sigma^2 = \int x^2 p(x)dx$ of a continuous variable $-\infty < x < \infty$. Assigning unit measure m(x) to unit range of x, the constrained maximisation of $S = -\int p(x)\log p(x)dx$ yields the Gaussian pdf

$$p(x) = (2\pi)^{-\frac{1}{2}}\sigma^{-1}\exp(-(x-\mu)^2/2\sigma^2) ,$$

with entropy $S = \log((2\pi e)^{\frac{1}{2}}\sigma)$.

An interesting generalisation is to observe autocorrelation coefficients

$$\langle A_k\rangle = \sum_j \langle x_j x_{j-k}\rangle = \sum_j \int d\underline{x}\; p(\underline{x})\; x_j x_{j-k}$$

of a discrete time-series $(\ldots, x_{-1}, x_0, x_1, x_2, \ldots)$ for some or all of the lags k. Again assuming constant measure in $\underline{x}$-space, the entropy of the pdf $p(\underline{x})$ of the time-series is

$$S = -\int d\underline{x}\; p(\underline{x}) \log p(\underline{x}) .$$

Maximising this over $p(\underline{x})$ under the autocorrelation and normalisation constraints gives

$$p(\underline{x}) \propto \exp(-\sum_{jk}\lambda_k x_j x_{j-k})$$

where the λ_k are the Lagrange multipliers or "potentials" of the corresponding constraints, chosen to make the constraints take their correct values.

It is immediately seen that the least committal pdf is Gaussian: this has followed from maximum entropy analysis as a straightforward consequence of the quadratic nature of the constraints, and need not be imposed as a separate assumption[22,23]. Had, for example, fourth moments been observed, the least committal pdf would have been the inverse exponential of a quartic.

The quadratic form $\sum\lambda_k x_j x_{j-k}$ can be diagonalised by a Fourier transform if the time-series is periodic (so that the matrix x_{j-k} is circulant), or by a related transform if x_{j-k} is Toeplitz. Thus

$$p(\underline{x}) \propto \exp(-\sum_\nu \Lambda_\nu |X_\nu|^2)$$

where X is the Fourier transform of the time-series samples x. Using this (normalised) pdf, the power-spectrum components are

$$\langle W_\nu\rangle = \int |X_\nu|^2\; p(\underline{x})\; d\underline{x} = 1/\Lambda_\nu$$

so that the entropy

$$S = -\int p(\underline{x}) \log p(\underline{x})\; d\underline{x} = \sum_\nu \log(\pi e/\Lambda_\nu)$$

becomes

$$S = \sum_\nu \log(\pi e\langle W_\nu\rangle) .$$

This is the Burg formula[24,25], derived directly by maximum entropy with no extra assumptions. It measures the entropy of a stochastic process producing individual samples x from the time-series. It can be used, or the underlying $-\int p(\underline{x}) \log p(\underline{x})\; d\underline{x}$ form can be used directly, for filling in gaps in a time-series or for extrapolation.

4. PHYSICAL ENTROPIES

The distribution of physical objects is often a statistical process involving a pdf which in turn has an entropy. A straightforward illustration concerns $p(x|\mu)$, the probability of obtaining an individual realisation of a system x, given its ensemble average $\mu = \langle x \rangle$. The entropy $S = -\sum p(x) \log(p(x)/m(x))$ represents the flexibility allowed to realisations of x within the ensemble-average constraint.

Classical objects: The distribution of quasars on the sky

Let x_i, an integer variable, represent the number of quasars in cell i of the sky. The appropriate measure m is the limiting degeneracy

$$m(x_i) = z^{x_i}/x_i!$$

of x_i objects distributed among z subdivisions of cell i, where z is large. This yields the familiar Poisson formula

$$p(x_i) = e^{-\mu_i} \mu_i^{x_i} / x_i!$$

The corresponding entropy is

$$S = -\sum_0^\infty p(x_i) \log(p(x_i)/m(x_i))$$

Combining pixels, the entropy simply adds, leading to a total entropy

$$S = M \log z - M \log M - M - M \sum_i p_i \log p_i$$

where $M = \sum \mu_i$ and $p_i = \mu_i/M$. The only term which depends on the spatial distribution p_i of the quasars is $-\sum p_i \log p_i$.

This classical application makes it clear that entropy is not measuring any physical property of quasars. Rather, entropy parameterises one's state of knowledge about their physical distribution. As observations of the x_i become more refined, the entropy will change.

Bosons: Photons

Similar analysis holds, save that there is now only one degenerate state for each $x_i = 0,1,2,\ldots$, so that $m(x_i) = 1$. Hence maximum entropy yields

$$p(x_i) = \left(\frac{1}{\mu_i+1}\right)\left(\frac{\mu_i}{\mu_i+1}\right)^{x_i}$$

with $S = (\mu_i+1) \log(\mu_i+1) - \mu_i \log \mu_i$

as in the gambling example above. This formula has two interesting limits. For large μ_i it becomes $S \sim \log \mu_i$. For small μ_i it becomes $S \sim -\mu_i \log\mu_i$.

These formulae are very similar to those derived by Kikuchi and Soffer[26] for photon degeneracy numbers, save that they needed to generalise to the case of z (greater than 1) degrees of freedom. If we do so, we merely alter the degeneracy $m(x_i)$ to be $m(x_i) = (z+x_i-1)!/x_i!(z-1)!$. For finite z, our probability distribution takes a form intermediate between the simple Bose and the pure classical Poisson process.

Fermions: Electrons

Again, similar analysis holds, save that there are now only two allowed states $x_i = 0,1$, each having unit measure $m(x_i) = 1$. Maximum entropy yields

$$p(0) = 1 - \mu_i , \qquad p(1) = \mu_i$$

for which the entropy is

$$S = - \mu_i \log \mu_i - (1-\mu_i) \log(1-\mu_i) .$$

Here μ_i is already a probability distribution in its own right, because $x_i = 0$ and 1 are mutually exclusive. The generalisation to finite z uses the Bernoulli form

$$m(x_i) = z!/x_i!(z-x_i)! .$$

Continuous objects

An interesting case occurs when x_i is a continuous variable with uniform measure m, which could be taken to be unity, in $[0,\infty)$. This is the same as the degenerate boson limit, with exponential probability distribution

$$p(x_i) = \mu_i^{-1} \exp(-x_i/\mu_i)$$

and a total entropy

$$S = \sum \log(e\mu_i m_i).$$

It is also the Shannon-Weaver result[18] as used in the Burg algorithm for time-series analysis.

We should emphasise that all these formulae have been simply derived from the fundamental Shannon entropy formula by Jaynes' method of maximum entropy[27]. In information-theoretic terms, they measure the expected number of bits needed to encode a particular pattern $\{x_i\}$ given the ensemble averages $\{\mu_i\}$. Equivalently, they measure the time-bandwidth product of the communication channel needed to transmit this information.

5. ENTROPY OF A POSITIVE, ADDITIVE IMAGE

When reconstructing an image, one is concerned with its configurational structure in the form of a sequence of positive numbers x_i $(i = 1,2,3,...)$ representing the signals (fluxes or equivalent quantities) in the pixels. The most direct way of identifying $\{x_i\}$ with a probability distribution is simply to remove its dimensionality by considering only the pattern of proportions $p_i = x_i/\sum x$ which describes the "shape" (but not the overall intensity scale) of the image. As long as the numbers are positive and additive (so that if pixels i and j were combined, $x_i + x_j$ is the signal one would assign to the combination), these proportions obey the Kolmogorov axioms and are themselves a probability distribution. Having identified the shape $\{p_i\}$ of an image with a probability distribution, the entropy naturally takes the simple Shannon/Jaynes form

$$S = -\sum_i p_i \log(p_i/m_i) .$$

For an ordinary optical image, p_i can be simply interpreted as the probability that the next photon to be radiated from the image would come from pixel i. The entropy corresponds to the uncertainty in this photon's location. Equivalently, it is the number of bits needed to encode its position, given the overall shape $\{p_i\}$. Maximising this entropy keeps one's options as open as possible about the next photon, and this automatically produces the most uniform, featureless image that is consistent with the constraining observations. It gives a maximally non-committal answer to the fundamental question "Where would the next photon come from?". We contend that this question is the precise formal definition of the problem of image reconstruction, lying at the very heart of the problem.

We must stress that our interpretation of an image as a photon pdf does not depend on the physical characteristics of photons: only one photon is being considered and it is meaningless to discuss its quantum degeneracy. The interpretation is equally applicable to questions such as "Where will the next electron arrive on a TV screen?", "Where will the next radioactive decay occur in a solid?" and "Where will the next quasar be discovered on the sky?". Distributions of photons, electrons, nuclei and quasars are each valid images, as are the sequences of numbers in computer memory or grains of silver on a photographic plate which are used to encode them. Since these physical manifestations are all interchangeable (at least in principle) it is entirely natural that they should all be assigned the same information content, irrespective of their internal physics.

The Shannon/Jaynes entropy is the configurational entropy and nothing but. It stands above such questions as the quantisation of the radiation field, with the consequent physics. The physical entropy tells one whether or not photons radiated from a given external object do actually obey Bose statistics: if they do, then the Bose physical entropy will attain its allowed maximum. The configuration entropy tells one about the shape of the external object: maximising it yields a reconstructed image which is maximally non-committal about the shape of the external object itself.

6. PHYSICAL ENTROPY REJECTED FOR IMAGE RECONSTRUCTION

In a significantly referenced paper, Kikuchi and Soffer[26] nevertheless recommended using the physical entropy of photons. As they pointed out, and as we saw above, this reduces to $-\sum p_i \log p_i$ for non-degenerate photons (which includes most of optical astronomy) and to $\sum \log x_i$ for degenerate photons (radio astronomy). Now the physical entropy of the photon arrival pattern measures the number of bits of information needed to encode the arrival pattern as a sequence of integers, given an ensemble-average constraint on the image. Although it does contain terms which depend on the shape of the image, it is really a design feature of a hypothetical experiment to measure photon occupation numbers in repeated observations of radiation from a given external object. It gives the maximum brightness the radiating object could have before it would be expected to saturate the communication channel from this experiment. As such, it has only a tenuous connection with image reconstruction.

We do, of course, entirely agree with Kikuchi and Soffer in their aim of "attempting to find the radiant spatial power pattern of the object". Our main criticism of their logic is that, by bringing in the duration of observation via a total number N of photons, they convert the power pattern (which is not physically quantised) into a photon pattern (which is quantised). Thus their definition of what they mean by an image involves some number N, proportional to the duration of the experiment and the intensity of the radiation, as well as the external power pattern itself. Since this number is included in their entropy, they fail to provide full objectivity: their entropy depends partly upon an accident of observation. We can restore objectivity by focussing on the location of the next photon to arrive. Since this does not involve how long one has to wait for it, the corresponding probability distribution faithfully reflects the spatial power pattern. In fact, it does not matter that the light is quantised at all, and we have returned to our direct identification of the shape of an image with a probability distribution.

There is a related criticism of the recommendation to use a form of entropy having different algebraic limits for degenerate and non-degenerate photons. Consider two telescopes X and Y. X observes an external object via a degenerate photon flux: the corresponding physical entropy is essentially $\sum \log x_i$. Y is an identical instrument having the same point-spread-function and observing an identical bandwidth, although ahead of it lies a uniform absorber of sufficient opacity K that the photons reaching Y are non-degenerate: the corresponding physical entropy is essentially $-\sum p_i \log p_i$. Y observes the external object for K times longer than X (thus increasing the number of degrees of freedom allowed to the observed photons by a factor K) so that the long-term datasets produced by the two telescopes are identical, in the sense of ensemble averages. Identical datasets, related to the same object via identical point-spread-functions, ought to be analysed in identical fashion. The physical entropy fails to be invariant under the signal-preserving transformation (intensity $\to$ intensity/K, time $\to$ K.time) and must be rejected for image reconstruction.

Because it has often been suggested that it makes very little difference what entropy or prior information is used to reconstruct an image[28,29], we now give two examples in which the entropy forms $-\sum p_i \log p_i$ and $\sum \log x_i$ give dramatically different results.

7. COMPARISON FOR MOMENT DATA

Suppose that we are given moment data

$$\int_{-\infty}^{\infty} t^n \, x(t)\, dt \;=\; \begin{cases} 1/(n+1) & (n \text{ even}) \\ 0 & (n \text{ odd}) \end{cases}$$

up to some maximum value N.

Suppose that we maximise the Shannon/Jaynes entropy under these constraints. Invoking a potential λ_n for each given moment yields

$$\delta\Big(-\int p(t) \log p(t)\, dt + \sum_n \lambda_n \int t^n \, x(t) dt \Big) = 0$$

so that

$$x(t) = \exp(-\, Q_N(t))$$

where $Q_N(t)$ is a polynomial of degree N. Direct numerical computation gives the results shown in Figure 1 for N = 4,8,12. As $N \to \infty$, the results converge uniformly (piecewise) to a square wave.

Suppose now that we maximise the physical entropy in the form $\sum \log x_i$, as suggested by Burg[24], by Kikuchi and Soffer[26] and their followers[30-33], and discussed for the radio sky by Ables[34]. Invoking potentials as before, we obtain

$$x(t) = 1/R_N(t)$$

where $R_N(t)$ is a polynomial of degree N. There is a small difficulty in that the highest (Nth) moment is divergent at infinity for any fixed $R_N(t)$, but this can be circumvented by restricting the range of integration to some large interval $[-T,T]$. The coefficient of t^N in R_N then decreases as T^{-1}. The character of the Burg solution is quite different from the smooth, uniformly convergent Shannon/Jaynes solution. Being the reciprocal of a polynomial, it exhibits N poles in conjugate pairs in the complex plane (if N is odd, the single real pole lies outside the interval $[-T,T]$). As we proceed to the limit $T \to \infty$, these poles squeeze the real axis, so that the solution approaches a set of $[N/2]$ delta functions. The positions are the zeros t_k of the Legendre polynomials $P_{[N/2]}(t)$ and their areas are the weights

$$w_k = 2/(1-t_k^2)P'_{[N/2]}(t_k)^2$$

conventionally used for Gaussian numerical integration[35]. This result follows from the fact that

$$\sum_k w_k q(t_k) = \int_{-1}^{1} q(t)\, dt$$

for any polynomial q of degree up to and including N. In particular, by taking

$$q(t) = \begin{cases} t^n/2 & |t| < 1 \\ 0 & |t| \geq 1 \end{cases}$$

we reach, for $n \leq N$,

$$\sum_k w_k t_k^n = \begin{cases} 1/(n+1) & \text{(n even)} \\ 0 & \text{(n odd)} \end{cases}$$

so that all given moments of the set of delta functions are correct, and we have the unique solution which fits the data and is the reciprocal of a polynomial of appropriate degree. Figure 2 illustrates solutions for N = 4,8,12. For large T, the heights and inverse widths of the spikes scale as the total width 2T of the interval, and are drawn correctly for T = 2.

This spiky behaviour is inherent in the $\sum \log x_i$ form, because the solution is always of the type

$$x(t) = 1/(\text{sum of constraints})$$

and the relevant sum will have zeros, normally near the real t-axis. Furthermore, the positions of the spikes are determined by the highest measured order. As more moments are measured, the number of spikes increases and their positions move. The solution for the $-\sum p_i \log p_i$ form is of the type

$$x(t) = \exp(-\text{sum of constraints})$$

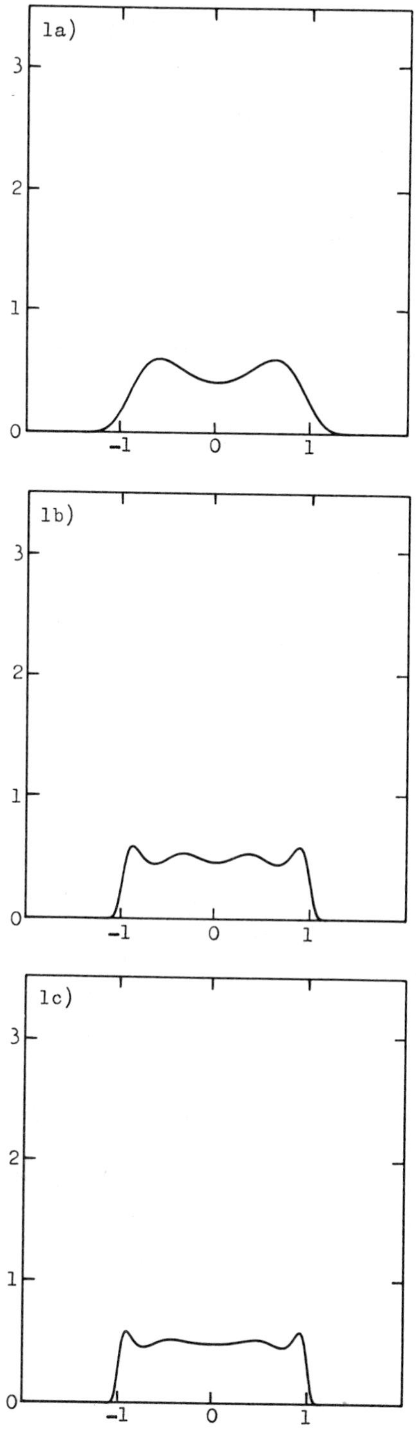

Fig. 1. Shannon/Jaynes reconstrustion of square wave from moment data.
a) N=4 b) N=8 c) N=12

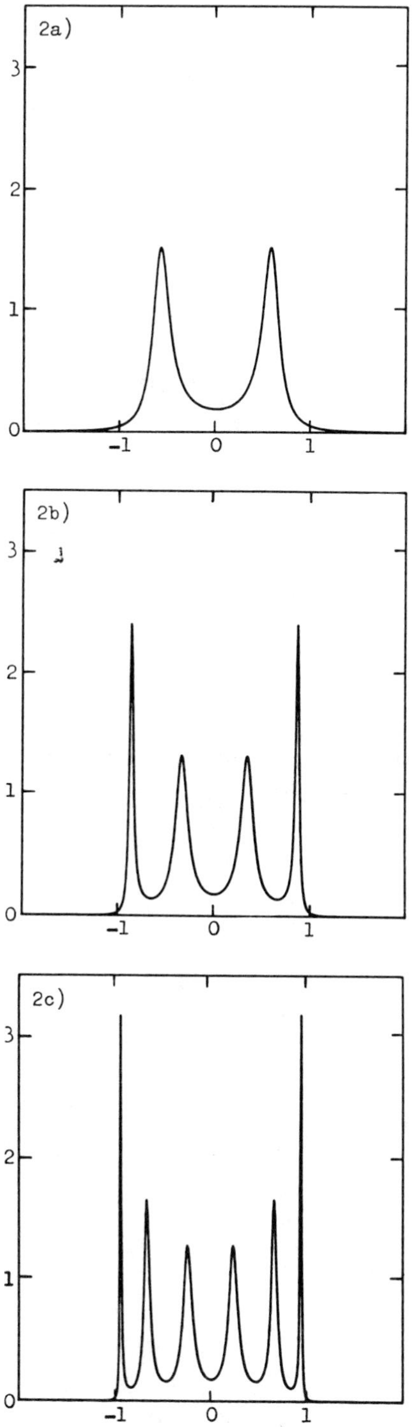

Fig.2. Burg reconstruction of square wave from moment data.
a) N=4 b) N=8 c) N=12

which is far less prone to produce atypically high values of x.

Had the moments taken values different from $1/(n+1)$, the solution would still have exhibited the same qualitative behaviour. Suppose we had taken moments appropriate to a positive function $\rho(t)$. Then the Shannon/Jaynes solution would have converged uniformly to $\rho(t)$ as $N \to \infty$. The Burg solution, on the other hand, would have been a set of delta functions at the zeros of the polynomials $S_{[N/2]}(t)$, where the $S_j(t)$ are the polynomials orthogonal over $\rho(t)$.

The positive function $\rho(t)$ could itself consist of a finite number r of spikes. The Shannon/Jaynes solution will converge uniformly to these, reaching exact agreement after observing $N = 2r$ moments because there is then only one feasible result. Figure 3 illustrates solutions for six equally spaced spikes

$$x(t) = \delta(x-5) + \delta(x-3) + \delta(x-1) + \delta(x+1) + \delta(x+3) + \delta(x+5)$$

$(r = 6)$ and $N = 4,8,12$. The Burg solution (Figure 4) reaches the same result via reconstructions consisting of increasing numbers of spikes in the wrong places. For large T, the heights and widths of the spikes scale as the total width $2T$ of the interval and are drawn correctly for $T = 10$. Clearly the Shannon/Jaynes solution, being less committal, is less misleading.

8. COMPARISON FOR THREE-DIMENSIONAL IMAGES

Suppose that we are given linear data

$$D_k = \int dV\, x(\xi,\eta,\zeta)\, R_k(\xi,\eta,\zeta)$$

related to a three-dimensional object $x(\underline{\xi})$ via response functions $R_k(\underline{\xi})$. This sort of application occurs in seismology, crystallography and medical body-scanning. A simple problem of this type concerns Fourier data on a unit cube of material

$$D_{hkl} = \int_{-\frac{1}{2}}^{\frac{1}{2}} dV\, x(\xi,\eta,\zeta) \cos 2\pi(h\xi + k\eta + l\zeta) .$$

We take the following particularly clear and powerful example from Nityananda and Narayan[29]. The data are

$$D_{000} = 1 , \quad D_{100} = D_{010} = D_{001} = 1/2 .$$

Suppose that we maximise the Shannon/Jaynes entropy under these constraints. Invoking potentials for the data yields

$$x(\underline{\xi}) = \exp(-\mu - \lambda_1 \cos 2\pi\xi - \lambda_2 \cos 2\pi\eta - \lambda_3 \cos 2\pi\zeta)$$

in which we may assign $\lambda_1 = \lambda_2 = \lambda_3$ because of symmetry. Fitting the numerical values of the data gives the well-behaved positive function

$$x(\underline{\xi}) = 0.391 \exp 1.161(\cos 2\pi\xi + \cos 2\pi\eta + \cos 2\pi\zeta)$$

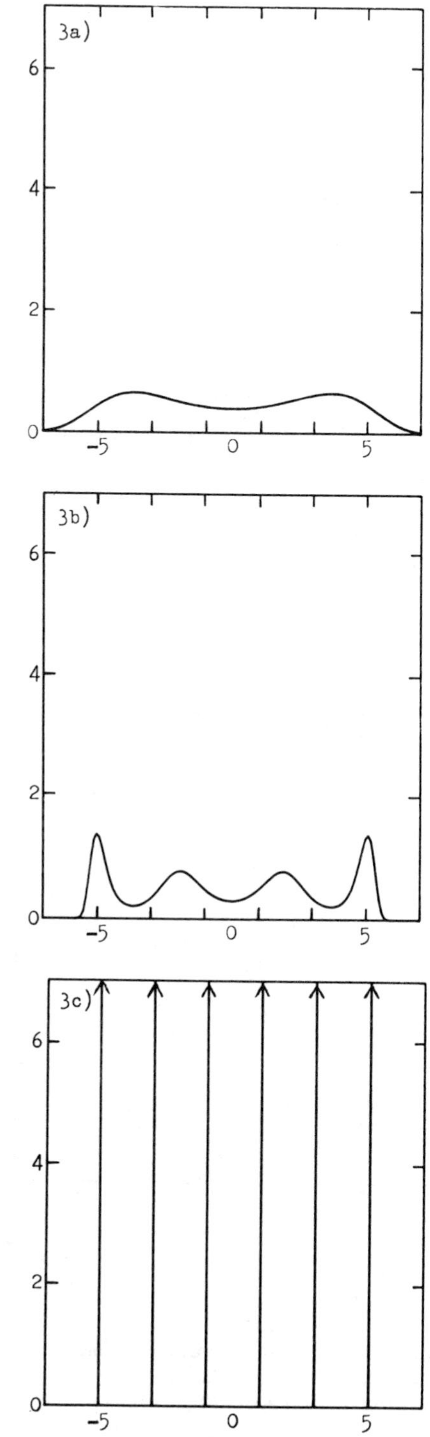

Fig.3. Shannon/Jaynes reconstruction of six spikes from moment data.
a) N=4 b) N=8 c) N=12

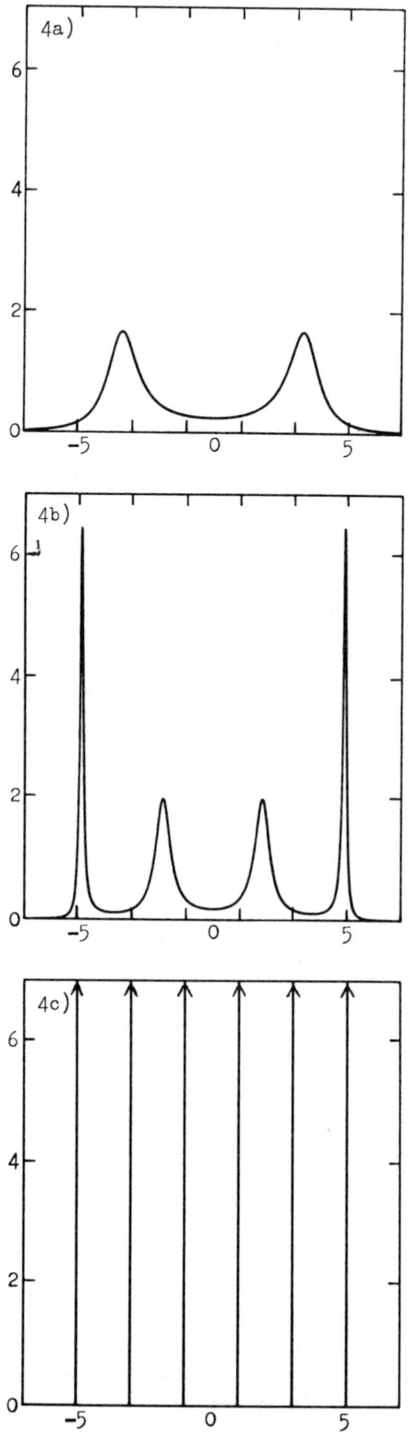

Fig.4. Burg reconstruction of six spikes from moment data.
a) N=4 b) N=8 c) N=12

shown in cross-section in Figure 5.

Suppose now that we maximise the entropy in the form $\sum \log x_i$. Invoking potentials, the variational equation gives

$$x(\underline{\xi}) = 1/(\mu + \lambda_1 \cos 2\pi\xi + \lambda_2 \cos 2\pi\eta + \lambda_3 \cos 2\pi\zeta)$$

with $\lambda_1 = \lambda_2 = \lambda_3$ by symmetry. A difficulty arises because this solution only remains positive everywhere if $\lambda < \mu/3$. At this limit, volume integrals of x remain finite and the "degree of structure" is $D_{100}/D_{000} = 0.34$. The variational solution does not allow any greater degree of structure than this, even though the actual data had $D_{100}/D_{000} = 0.5$, and we have already seen that this is perfectly consistent with positivity.

The only way of keeping $x > 0$ everywhere and truly maximising $\int dV \log x$ is to allow part of the solution to condense into delta functions, at which the variational approach breaks down. The solution is

$$x(\underline{\xi}) = 0.242\ \delta(\underline{\xi}) + 1.5/(3 - \cos 2\pi\xi - \cos 2\pi\eta - \cos 2\pi\zeta)$$

shown in cross-section in Figure 6. Almost a quarter of the total reconstruction is forced into a single point, without any evidence for this in the data.

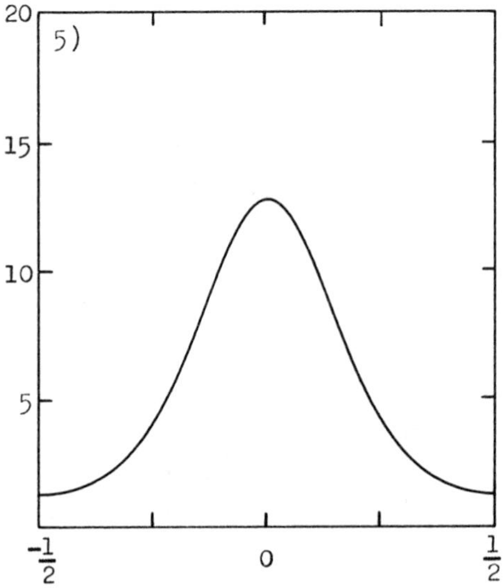

Fig.5. Shannon/Jaynes reconstruction from three-dimensional Fourier data (section along x-axis).

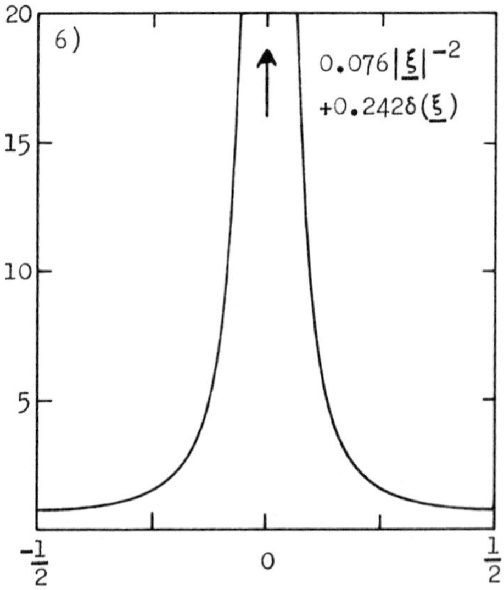

Fig.6. Burg reconstruction from three-dimensional Fourier data (section along x-axis).

Delta functions do not, in fact, contribute to the Burg formula at all, because of the logarithm. By contrast, they give infinite contribution to the Shannon/Jaynes formula, so that Shannon/Jaynes reconstructions only give delta functions if the data categorically demand them.

As well as giving this particular Fourier example, Nityananda and Narayan pointed out that quite generally in three (or more) dimensions, the reciprocal of a band-limited function

$$x(\underline{\xi}) = 1/\sum \lambda_k R_k(\underline{\xi})$$

may not be able to fit a set of data, even though the data are consistent with smooth and positive (though substantially non-uniform) reconstructions. At the limiting degree of structure, the denominator $\sum\lambda_k R_k(\underline{\xi})$ will normally go to zero as $0(|\delta\underline{\xi}|^2)$ for twice-differentiable response functions R_k, at some point or points $\underline{\xi}$. Although the reconstruction x becomes infinite there as $0(|\delta\underline{\xi}|^{-2})$, volume integrals over this point remain finite because of geometrical factors, so that the fit to the data D_k remains well-behaved. Beyond the limiting degree of structure, one is forced to add delta functions to the existing singularities of the reconstruction.

We have given these examples to illustrate dramatically the difference which can be made by changing the type of the entropy being maximised. The Shannon/Jaynes solution is non-committal about the position of ξ from the distribution $x(\xi)$, and it gives smooth, uniformly convergent reconstructions x. The Burg solution is maximally non-committal about samples $x(\xi)$ from an ensemble-average constrained distribution, which involves maximising the hypervolume beneath the solution $\langle x(\xi)\rangle$. It makes the solution as large as possible.

9. GENERALISATIONS AND DEVELOPMENTS

The direct identification of an image with a probability distribution can be generalised in several ways, which allow further developments of maximum entropy data analysis. We give three examples from current research at Cambridge.

<u>2-Dimensional reconstructions from 1-dimensional data</u>

How can one reconstruct a two dimensional density function

$$x(\omega,k), \quad \omega = \text{frequency}, \quad k = \text{decay}$$

from one-dimensional time-series data

$$D(t) = \iint x(\omega,k)\cos\omega t\, e^{-kt}\, d\omega\, dk \qquad (t > 0) \quad ?$$

Sibisi[36] has investigated this problem with applications to spectroscopy in mind, and this section reports his work. In chemical nuclear-magnetic-resonance spectroscopy, x represents the number density of nuclei having gyrofrequency

ω and relaxation rate k in a sample, and $D(t)$ represents the net magnetic moment of the sample as the nuclei precess in an external magnetic field. Detailed chemical information can be aquired from such spectra, since the gyrofrequency of each nucleus is sensitive to its local chemical environment.

A simulation consisting of five species of nuclei, each with different frequency and decay (Figure 7) produces the data shown in Figure 8. Clearly this is a set of decaying oscillations, and Figure 9 plots its inverse Fourier transform. This shows five peaks, though the right-most pair are barely resolved, and the decays have resulted in considerable line-broadening, different for each line. To some extent line broadening can be reduced by apodisation (pre-multiplying the data by a function like $\exp(+kt)$ to correct for an average decay rate), but this still leaves imperfect profiles for each individual line. Figure 10 shows such a spectrum, in which the left-most peak is over-resolved and "ringing" while the right-most peaks are under-resolved and broad. Also noise has been amplified. Conventional analysis gives spectra of this type.

To apply maximum entropy, one normalises the desired function $x(\omega,k)$ to a pdf, corresponding to the fundamental question "Where (in the ωk plane) would the next nucleus be?" Maximising the entropy of this pdf gives a maximally noncommittal distribution of nuclei in frequency and decay simultaneously. Taking x to be normalised, the entropy is

$$S = -\iint x(\omega,k) \log(x(\omega,k)/m(\omega,k)) \, d\omega \, dk .$$

In spite of the severe non-uniqueness inherent in this formalisation, the question being asked does nevertheless correspond to what one actually wants to know.

An interesting twist to this problem is that different (ω,k) points may not be equivalent. Indeed Sibisi argues that the natural measure in the frequency/decay time phase-plane (ω,T) is uniform, since ω is translation invariant and ω and T have inverse dimensions. This transforms to

$$m(\omega,k) = \partial(\omega,T)/\partial(\omega,k) = k^{-2}$$

in the (ω,k) plane.

With this measure, the 2-dimensional maximum entropy solution for x is shown in Figure 11. Resolution in the decay direction is poor: recovery of decay information is a notoriously ill-posed problem. Resolution in the frequency direction shows a spectacular improvement over conventional spectra. Figure 12 shows the projection $\int x \, dk$ of the full reconstruction, in which all five peaks are clearly resolved and noise suppression is excellent.

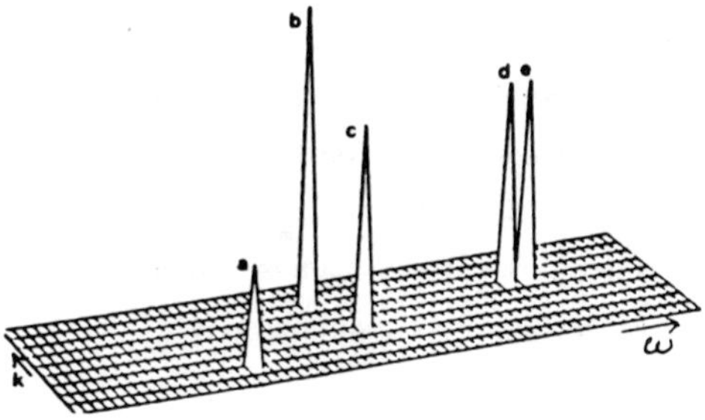

Fig.7 Simulation

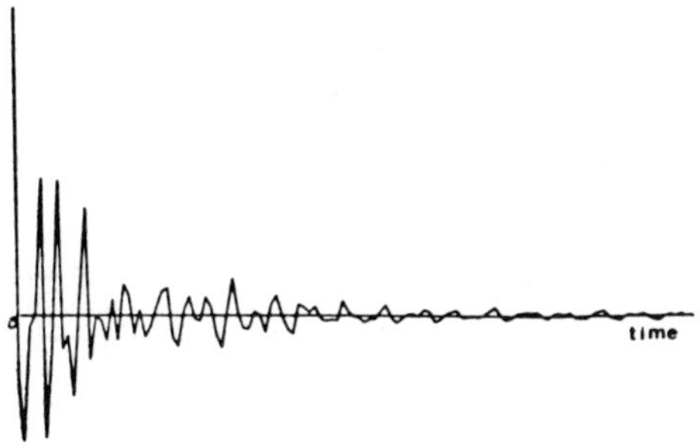

Fig. 8 Corresponding time-series data

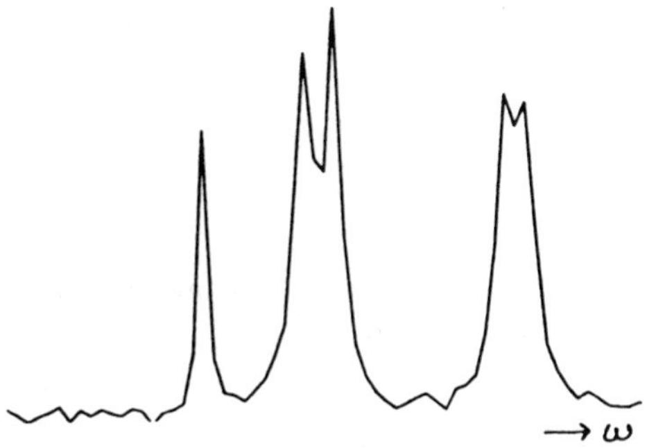

Fig. 9 Conventional reconstruction. Fourier transform of time-series.

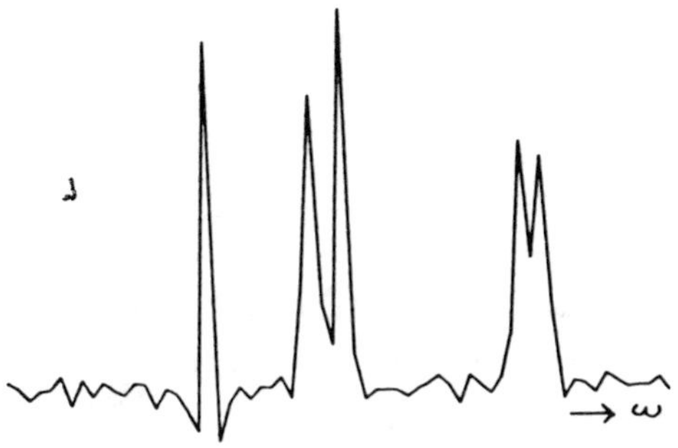

Fig. 10 Fourier transform of apodised time-series

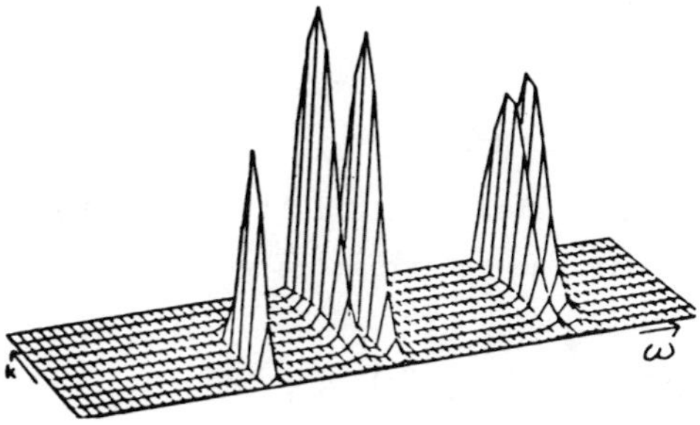

Fig. 11 MEM 2-dimensional reconstruction

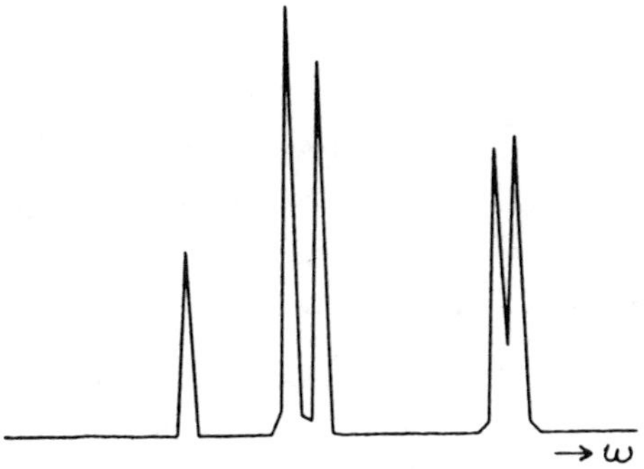

Fig. 12 Projection of Fig. 11 to frequency axis

Polarisation data

How can one interpret observations of polarised radiation from an object? The "next photon" may come from any pixel, and also may be polarised L.H. or R.H. or any combination, so that a polarisation image must display more than just intensity.

Partly polarised radiation is characterised by the four Stokes' parameters I,Q,U,V per pixel. In the quantum mechanical circular representation

$$\left|\begin{matrix}1\\0\end{matrix}\right\rangle = \text{L.H.} \qquad \left|\begin{matrix}0\\1\end{matrix}\right\rangle = \text{R.H.}\ ,$$

we have

$$I = \begin{pmatrix}1 & 0\\0 & 1\end{pmatrix} = \text{total intensity,} \qquad V = \begin{pmatrix}1 & 0\\0 & -1\end{pmatrix} = \text{net left-hand,}$$

$$Q = \begin{pmatrix}0 & 1\\1 & 0\end{pmatrix} = \text{net x-linear,} \qquad U = \begin{pmatrix}0 & i\\-i & 0\end{pmatrix} = \text{net quadrature.}$$

Observations of I,Q,U,V fix the density matrix

$$\rho = \begin{pmatrix}I+V & Q-iU\\Q+iU & I-V\end{pmatrix} /I$$

This density matrix is the quantum mechanical generalisation of the probability distribution function, satisfying $tr(\rho) = 1$ and having entropy $S = tr(-\rho \log \rho)$.

Only in exceptional cases will the polarisation state be pure Q or U or V. More usually, it will be some partially polarised combination, best expressed in terms of the diagonal representation of the density matrix. This diagonal representation is

$$\rho = \begin{pmatrix}(1+\alpha)/2 & 0\\0 & (1-\alpha)/2\end{pmatrix}$$

where $\alpha = (Q^2 + U^2 + V^2)^{\frac{1}{2}}/I$

is the overall degree of polarisation, ranging from 0 to 1. The eigenvalues $(1 \pm \alpha)/2$ are the probabilities that the "next photon" would fall into either eigenstate. Correspondingly, the uncertainty in the polarisation state is given by the entropy

$$S^{pol} = tr(-\rho \log \rho) = -\frac{1+\alpha}{2}\log\frac{1+\alpha}{2} - \frac{1-\alpha}{2}\log\frac{1-\alpha}{2}.$$

Completely polarised radiation $(\alpha = 1)$ has $S^{pol} = 0$ because the polarisation state is fully predictable, whilst unpolarised radiation $(\alpha = 0)$ has $S^{pol} = \log 2 = 1$ bit. With more than one pixel in the image, the density matrix diagonalises to

$$\rho = \text{diag}\ (p_1(1+\alpha_1)\ ,\ p_1(1-\alpha_1)\ ,\ p_2(1+\alpha_2)\ ,\ p_2(1-\alpha_2)\ ,\ \ldots\ )\ /2$$

where $p_j = I_j/\sum I$. The entropy

$$S = -\sum p_j(\log p_j + S_j^{pol})$$

measures the uncertainty in the answer to the question "What is the density matrix of the next photon to arrive?".

The alternative form

$$\begin{aligned} S &= \mathrm{tr}(\log I\rho) \\ &= \sum \log(I_j^2 - U_j^2 - V_j^2 - Q_j^2) \end{aligned}$$

has also been proposed[37,38]. That form is the physical entropy of the wave field constrained by ensemble-average measurements: it is the correct solution to the different problem of estimating individual realisations of polarisation occupation numbers.

Therefore, for the reasons given above one should use the Shannon/Jaynes form

$$S = -\sum p_j(\log p_j + S_j^{pol}) \ ,$$

when reconstructing a general-purpose image from polarisation data.

Blind deconvolution

How can one interpret a blurred picture $D_k = \sum f_j b_{k-j}$ without knowing the point-spread-function b?

In ordinary (incoherent) optics, f_j is positive and can be identified with a pdf having entropy

$$S_f = -\sum p_j \log p_j \ , \quad p_j = f_j/\sum f \ .$$

Likewise b_k is positive and can be identified with the pdf associated with the question "Where would the next photon arrive in a blurred picture of a point object?". Its entropy is

$$S_b = -\sum q_k \log q_k \ , \quad q_k = b_k/\sum b \ .$$

Entropy is additive over pdf's in different spaces, so that one's total ignorance is expressed as some linear combination

$$\theta S_f + (1-\theta)S_b \ , \quad 0 < \theta < 1$$

of the entropies of the image and of the point-spread-function. The following argument suggests a value of θ. In general, exp(entropy) defines the effective number of independent choices allowed by a pdf, so that samples of f and b are effectively distributed over $\exp S_f$ and $\exp S_b$ pixels respectively. Juxtaposing these spaces into one combined system, a single random sample would be drawn from f and b in proportion $\theta/(1-\theta) = \exp S_f/\exp S_b$. Thus the expectation entropy is $\theta S_f + (1-\theta)S_b$, as in a two-phase physical system, and this is taken to represent one's total degree of ignorance.

Of course there are inevitable ambiguities in blind deconvolution. For example, a sharp picture of a brick wall could equally well be described as a pattern of points blurred by a brick shaped point-spread-function. Nevertheless results obtained with this algorithm are encouraging (Newton, private communication).

10. CONCLUSIONS

The shape of any positive, additive image can be directly identified with a probability distribution p. Accordingly, whether it be for radio astronomy, optical or X-ray astronomy, or for reconstruction of any other positive, additive image, or even for spectral analysis of time-series, the configurational entropy is

$$S = -\sum_i p_i \log(p_i/m_i).$$

Maximum entropy in the form we commend yields reconstructions which are maximally non-committal answers to the fundamental question "Where would the next photon come from?". This relates directly and unequivocally to the shape of the object being observed.

Maximum entropy can also be applied to other problems such as time-series analysis and particle distributions, and formulae such as the Burg formula can be derived. Whilst perfectly valid in context, these are not applicable to image reconstruction. The Burg formula, for instance, was designed to predict samples from a pdf, not to reconstruct the shape of the pdf. The shape it produces is a means to an end, not the end itself, so it is hardly surprising that it may exhibit peculiarities. If, for particular types of data such as autocorrelation coefficients or one- or two-dimensional Fourier transform components, the resulting shapes are also reasonably smooth and convergent, that is merely a fortunate accident.

Finally, the direct Shannon/Jaynes approach is capable of extensive generalisation. It allows the development of a whole new range of powerful and practical techniques of data analysis.

11. ACKNOWLEDGEMENTS

We thank Dr. Geoff Daniell and the increasing number of official and unofficial members of the maximum entropy group at Cambridge for many stimulating discussions. Ed Jaynes, of course, first expounded these ideas in his brilliantly clear papers[20,39-44].

BIBLIOGRAPHY

1. Gull, S.F., Daniell, G.J., "Image reconstruction from incomplete and noisy data", Nature, 272 (1978), 686-690.

2. Frieden, B.R., "Restoring with maximum likelihood and maximum entropy", J. Opt. Soc. Am. 62 (1972), 511-518.

3. Frieden, B.R., Burke, J.J., "Restoring with maximum entropy. II: Superresolution of photographs of diffraction-blurred impulses", J. Opt. Soc. Am. 62 (1972), 1202-1210.

4. Frieden, B.R., Wells, D.C., "Restoring with maximum entropy. III: Poisson sources and background", J. Opt. Soc. Am. 68 (1978), 93-103.

5. Skilling, J., Strong, A.W., Bennett, K., "Maximum entropy image processing in gamma-ray astronomy", Mon. Not. R. Astr. Soc. 187 (1979), 145-152.

6. Frieden, B.R., "Statistical models for the image restoration problem", Comp. Graphics Image Processing 12 (1980), 40-59.

7. Daniell, G.J., Gull, S.F., "Maximum entropy algorithm applied to image enhancement", IEE Proc. 127E (1980), 170-172.

8. Bryan, R.K., Skilling, J., "Deconvolution by maximum entropy as illustrated by application to the jet of M87", Mon. Not. R. Astr. Soc. 191 (1980), 69-79.

9. Fabian, A.C., Willingale, R., Pye, J.P., Murray, S.S., Fabbiano, G., "The X-ray structure and mass of the Cassiopeia A supernova remnant", Mon. Not. R. Astr. Soc. 193 (1980), 175-188.

10. Willingale, R., "Use of the maximum entropy method in X-ray astronomy", Mon. Not. R. Astr. Soc. 194 (1981), 359-364.

11. Burch, S.F., Gull, S.F., Skilling, J., "Image restoration by a powerful maximum entropy method", Comp. Vision Graphics Image Processing (1983), (in press).

12. Minerbo, G., "MENT: A maximum entropy algorithm for reconstructing a source from projection data", Comp. Graphics Image Processing 10 (1979), 48-68.

13. Kemp, M.C., "Maximum entropy reconstructions in emission tomography", Medical Radionuclide Imaging 1 (1980), 313-323.

14. Collins, D.M., "Electron density images from imperfect data by iterative entropy maximisation", Nature 298 (1982), 49-51.

15. Skilling, J., "Maximum entropy image reconstruction from phaseless Fourier data". Paper presented at Opt. Soc. Am. meeting on Signal recovery and synthesis with incomplete information and partial constraints, Incline Village, Nevada (1983).

16. Burch, S.F., "Comparison of image generation methods", UKAEA Harwell report AERE-R 9671 (1980).

17. Shannon, C.E., "A mathematical theory of communication", Bell System Tech. J. 27 (1948), 379-423 and 623-656.

18. Shannon, C.E., Weaver, W., The mathematical theory of communication, Univ. Illinois, Urbana, 1949.

19. Ash, R.B., Information theory, Interscience, N.Y., 1965, pp.5-12.

20. Jaynes, E.T., "Prior probabilities", IEEE Trans. SSC-4 (1968), 227-241.

21. Shore, J.E., Johnson, R.W., "Axiomatic derivation of maximum entropy and the principle of minimum cross-entropy", IEEE Trans. IT-26 (1980), 26-37.

22. Ulrych, T.J., Bishop, T.M., "Maximum entropy spectral analysis and autoregressive decomposition", Rev. Geophys. Space Phys. 13 (1975), 183-200.

23. Ulrych, T.J., Clayton, R.W., "Time-series modelling and maximum entropy", Phys. Earth and Plan. Interiors 12 (1976), 188-200.

24. Burg, J.P., "Maximum entropy spectral analysis", Paper presented at 37th meeting of the Society of Exploration Geophysicists, Oklahoma City (1967).

25. Burg, J.P., "The relationship between maximum entropy spectra and maximum likelihood spectra", Geophysics 38 (1972), 375-376.

26. Kikuchi, R., Soffer, B.H., "Maximum entropy image restoration. I: The entropy expression", J. Opt. Soc. Am. 67 (1977), 1656-1665.

27. Grandy, W.T., "Indistinguishability, symmetrisation and maximum entropy", Eur. J. Phys. 2 (1981), 86-90.

28. Hogbom, J.A., "The introduction of a priori knowledge in certain processing algorithms", in Image formation from coherence functions in astronomy, Gronigen (ed. D. Reidel) (1978).

29. Nityananda, R., Narayan, R., "Maximum entropy image reconstruction - a practical noninformation theoretic approach", J. Astrophys. Astron. (1983) (in press).

30. Wernecke, S.J., "Two-dimensional maximum entropy reconstruction of radio brightness", Radio Science 12 (1977), 831-844.

31. Wernecke, S.J., d'Addario, L.R., "Maximum entropy image reconstruction" IEEE Trans. C-26 (1977), 351-364.

32. Edward, J.A., Fitelson, M.M., "Notes on maximum entropy processing", IEEE Trans. IT-19 (1973), 232-234.

33. Newman, W.I., "A new method of multidimensional power spectral analysis", Astron. Astrophys. 54 (1977), 369-380.

34. Ables, J.G., "Maximum entropy spectral analysis", Astron. Astrophys. Suppl. 15 (1974), 383-393.

35. Abramowitz, M., Stegun, I.A., Handbook of mathematical functions, Dover, N.Y. (1970), p. 916.

36. Sibisi, S., "Two-dimensional reconstruction from one-dimensional data by maximum entropy", Nature 301 (1983), 134-136.

37. Ponsonby, J.E.B., "An entropy measure for partially polarised radiation", Mon. Not. R. Astr. Soc. 163 (1977), 369-380.

38. Nityananda, R., Narayan, R., "Reconstruction of a polarised brightness distribution by the maximum entropy method", Astron. Astrophys. 118 (1982), 194-196.

39. Jaynes, E.T., "Information theory and statistical mechanics", Phys. Rev. 106 (1957), 620-630.

40. Jaynes, E.T., "Information theory and statistical mechanics II", Phys. Rev. 108 (1957), 171-190.

41. Jaynes, E.T., "Information theory and statistical mechanics" in Brandeis 1962 Summer Institute in Theoretical Physics, ed. K. Ford, Benjamin, N.Y., 1963.

42. Jaynes, E.T. "The well-posed problem", Foundation of Physics 3 (1973), 477-492.

43. Jaynes, E.T. "Confidence intervals vs. Bayesian intervals", in Foundations of Probability Theory, Statistical Inference, and Statistical Theories of Science, ed. W.L. Harper & C.A. Hooker, D. Reidel (1976),pp.175-257.

44. Jaynes, E.T. "Where do we stand on Maximum Entropy," in The Maximum Entropy Formalism, ed. R.D. Levine & M. Tribus, MIT Press, Cambridge Mass. (1978) pp.15-118.

DEPT. OF APPLIED MATHEMATICS AND
THEORETICAL PHYSICS
CAMBRIDGE UNIVERSITY
SILVER STREET
CAMBRIDGE CB3 9EW
ENGLAND

MULLARD RADIO ASTRONOMY OBSERVATORY
CAVENDISH LABORATORY
CAMBRIDGE UNIVERSITY
MADINGLEY ROAD
CAMBRIDGE CB3 OHE
ENGLAND

ABCDEFGHIJ-CM-8987654